NYC

Department of Education

BUREAU OF NONPUBLIC SCHOOL REIMBURSABLE SERVICES

Title I
High School Activities

with **THE GEOMETER'S**
SKETCHPAD®
VERSION 5

Key Curriculum Press
INNOVATORS IN MATHEMATICS EDUCATION

Writers:	Dan Bennett, Steven Chanan, Paul Kunkel, Lyublinskaya, Daniel Scher, Scott Steketee
Contributing Writers:	Masha Albrecht, Eric Bergofsky, Lyubomir Detchkov, Ned Diamond, Daniel Dudley, Allan Bergmann Jensen, Eric Kamischke, Jennifer North Morris, Ralph Pantozzi, Brooke Precil, Nathalie Sinclair, Kevin Thompson
Reviewers:	Pat Brewster, Steven Chanan, Christopher David, Molly Jones, Dan Lufkin, Aaron Madrigal, Marsha Sanders-Leigh, Daniel Scher, Scott Steketee, John Threlkeld, Bill Zahner, Danny Zhu
Editors:	Scott Steketee, Steven Chanan, Elizabeth DeCarli, Karen Greenhaus
Contributing Editors:	Josephine Noah, Cindy Clements, Sharon Taylor
Production Editors:	Angela Chen, Andrew Jones, Christine Osborne
Other Contributers:	Judy Anderson, Jason Luz, Marilyn Perry
Copyeditor:	Jill Pellarin
Printer:	Lightning Source, Inc.
Executive Editor:	Josephine Noah
Publisher:	Steven Rasmussen

Key Curriculum Press
1150 65th Street
Emeryville, CA 94608
510-595-7000
editorial@keypress.com
www.keypress.com

ISBN: 978-1-60440-246-9
10 9 8 7 6 5 4 3 2 1   15 14 13 12 11

Contents

Chapter 6: Constructions and Loci (Geometry)

Chapter 7: Triangle Properties (Geometry)

Chapter 8: Circles and Transformations (Geometry)

New York City Title I High School Activities
© 2012 Key Curriculum Press

Chapter 9: Exponents, Radicals, and Complex Numbers (Algebra II and Trigonometry)

Chapter 10: Relations and Functions (Algebra II and Trigonometry)

Chapter 11: Trigonometry and Statistics (Algebra II and Trigonometry)

Sketchpad Resources

Sketchpad Learning Center

The Learning Center provides a variety of resources to help you learn how to use Sketchpad, including overview and classroom videos, tutorials, Sketchpad Tips, sample activities, and links to online resources. You can access the Learning Center through Sketchpad's start-up screen or through the Help menu.

The Learning Center has three main sections:

Welcome Videos

These videos introduce Sketchpad from the point of view of students and teachers, and give an overview of the big ideas and new features of Sketchpad 5.

Using Sketchpad

This section includes 12 self-guided tutorials with embedded videos, 70 Sketchpad Tips, and links to local and online resources.

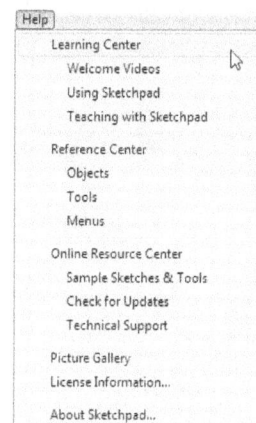

Teaching with Sketchpad

This section includes videos and articles describing how teachers make effective use of Sketchpad and how it affects their students' attitudes and mathematical understanding. There are over 40 sample activities, each with an overview, teaching notes, student worksheet, and sketches, that you can use with students to support your subject area, level, and curriculum.

Other Sketchpad Resources

Sketchpad contains resources for beginning and advanced users.

- **Reference Center:** This digital resource, which is accessed through the Help menu, is the complete reference manual for Sketchpad, with detailed information on every object, tool, and menu command. The Reference Center includes a number of How-To sections, an index, and full-text search capability.

- **Online Resource Center:** The Geometer's Sketchpad Resource Center (www.dynamicgeometry.com) contains many sample sketches and advanced toolkits, links to other Sketchpad sites, technical information (including updates and frequently asked questions), and detailed documentation for JavaSketchpad, which allows you to embed dynamic constructions in a web page.

- **Sketch Exchange:** The Sketchpad Sketch Exchange™ (sketchexchange.keypress.com) is a community site where teachers share sketches and other resources with Sketchpad users. Browse by keyword or topic for sketches that interest you, or ask questions and share ideas in the forum.

- **Sample Sketches & Tools:** You can access many sketches, including some with custom tools, through Sketchpad's Help menu. You can use some sample sketches as demonstrations, others to get tips and information about particular constructions, and others to access custom tools that you can use to perform special constructions. These sketches are also available under General Resources at the Sketchpad Resource Center (www.dynamicgeometry.com).

- **Online Courses:** Key Curriculum Press offers moderated online courses that last six weeks, allowing you to immerse yourself in learning how to use Sketchpad in your teaching. For more information, see Sketchpad's Learning Center, or go to www.keypress.com/onlinecourses.

- **Other Professional Development:** Key Curriculum Press offers free webinars on a regular basis. You can also arrange for one-day or three-day face-to-face workshops for your district or school. For more information, go to www.keypress.com/pd.

Addressing Grade-Level Learning Objectives

The table below shows how the activities in this collection align to the Title I Mathematics Learning Objectives for Algebra, Geometry, and Algebra II and Trigonometry. The following standards are not included in the table because they are addressed in many of the activities:

- Uses a variety of strategies to understand, represent, and solve problems

- Articulates problem solving processes and connects mathematical ideas to problem situations outside of mathematics

Some activities include pre-made sketches with dynamic models and representations that allow students to explore mathematical concepts using a variety of strategies. In other activities, students create their own sketches to understand, represent, and solve problems. The Student Worksheets pose questions that require students to articulate their mathematical processes. Some activities model real-world situations, connecting mathematics to problem situations outside of mathematics.

Activity Title	Learning Objective
Chapter 1: Operations on Numbers and Polynomials (Algebra)	
Adding Integers	Identifies and applies the properties of real numbers
Subtracting Integers	Performs operations with real numbers
Dividing Real Numbers	
Ratio and Proportion	Solves problems involving ratio and proportion
The Product of Two Binomials	Performs operations with polynomials
Squaring Binomials	
Chapter 2: Solving and Graphing Functions and Inequalities (Algebra)	
Solving Linear Equations by Balancing	Identifies, solves, and graphs functions and inequalities
Solving Inequalities by Balancing	
Graphing Inequalities in Two Variables	
Graphing Systems of Inequalities	
Absolute Value Functions	
Exponential Functions	
Changing Quadratic Function Forms	
Chapter 3: Geometric Properties (Algebra)	
The Pythagorean Theorem	Solves problems involving trigonometric ratios and the Pythagorean Theorem
Trigonometric Ratios	
Slopes of Parallel and Perpendicular Lines	Understands and applies properties of parallel and perpendicular lines

Activity Title	Learning Objective
Chapter 3: Geometric Properties (Algebra), *Continued*	
The Distributive Property: A Painting Dilemma A Rectangle with Maximum Area Direct Variation	Solves problems involving area, perimeter, and volume of two-dimensional shapes and three-dimensional figures
Chapter 4: Measurement, Data, and Probability (Algebra)	
Rates and Ratios	Solves problems involving ratio and proportion Calculates rates and converts between systems of measurement
Box and Whiskers Lines of Fit	Analyze, evaluates, and represents data
Wait for a Date	Solves problems involving probability
Chapter 5: Lines and Polygons (Geometry)	
Introducing Points, Segments, Rays, and Lines	Understands and applies the properties of lines, polygons, planes, and solids
Triangle Sum Exterior Angles in a Triangle Exterior Angles in a Polygon Polygon Angle Measure Sums	Applies the properties of interior and exterior angles of polygons
Different Slopes: The Slope of a Line Midsegments of a Trapezoid and a Triangle Midpoint Quadrilaterals	Applies and investigates the properties of slope, midpoint and distance formulas to line segments
Chapter 6: Constructions and Loci (Geometry)	
Constructing Isosceles Triangles Constructing Parallelograms Constructing Rectangles Constructing Rhombuses Constructing Isosceles Trapezoids	Uses a straightedge and compass to construct geometric figures
Parabolas: A Geometric Approach From Locus to Graph	Understands and applies basic and complex locus theorems
Chapter 7: Triangle Properties (Geometry)	
Triangle Congruence Similar Triangles—AA Similarity Similar Triangles—SSS, SAS, SSA	Uses theorems and/or postulates to prove that triangles are congruent or similar

Activity Title	Learning Objective
Chapter 7: Triangle Properties (Geometry), *Continued*	
Constructing Squares on a Triangle: The Pythagorean Theorem The Isosceles Right Triangle The 30°-60° Right Triangle	Applies properties of the Pythagorean Theorem and special triangles
Chapter 8: Circles and Transformations (Geometry)	
Chords in a Circle Tangents to a Circle Tangent Segments Arcs and Angles	Investigates, justifies, and applies theorems regarding chords, arcs, tangents and secants drawn to a circle
Properties of Reflection Reflections in the Coordinate Plane Translations in the Coordinate Plane Reflections across Two Parallel Lines Reflections across Two Intersecting Lines Glide Reflections Symmetry in Regular Polygons	Solves transformation problems involving: rotation, reflection, translation, and glide reflection
Chapter 9: Exponents, Radicals, and Complex Numbers (Algebra II and Trigonometry)	
Zero and Negative Exponents The Square Root Spiral	Performs operations on radicals and exponents
Multiplication of Complex Numbers In Search of Buried Treasure Powers of Complex Numbers	Performs arithmetic operations on complex numbers
Chapter 10: Relations and Functions (Algebra II and Trigonometry)	
Direct Variation Inverse Variation	Uses direct and inverse variation to solve for unknown values
Modeling Projectile Motion Exponential Functions Logarithmic Functions	Understands and applies exponential, logarithmic, and quadratic functions
Domain and Range Introducing Dynagraphs Relations and Functions	Defines a relation and a function, and determines its domain and range

Activity Title	Learning Objective
Chapter 11: Trigonometry and Statistics (Algebra II and Trigonometry)	
Right Triangle Functions	Understands and applies basic trigonometric functions
Radian Measure	Converts between radian and degree measures
Unit Circle and Right Triangle Functions Six Circular Functions	Understands and applies basic trigonometric functions
Normal Distribution	Interprets statistical studies, central tendency, and dispersion
Permutation and Combination	Calculates the number of possible permutations ($_nP_r$) and the number of possible combinations ($_nC_r$)
Fitting Functions to Data	Interprets statistical studies, central tendency, and dispersion

1

Operations on Numbers and Polynomials (Algebra)

Adding Integers

Use this activity as an introduction to integer addition for pre-algebra students, as a start-of-the-year refresher for Algebra 1 students, or as a supplemental activity for any student having difficulty with the topic. It's important for students to have a mental image of operations on integers. Even strong students who rely on verbal rules make careless mistakes that could be avoided by having an internalized picture.

The picture of addition presented here is a geometric model in which each number is represented by a vector. (The activity calls them *arrows* because students may not be familiar with the term *vector*.) Vectors incorporate both magnitude and direction (representing the absolute value and the sign of the integer), so practice with this model helps students understand how the signs of the addends come into play.

This activity contains lots of questions for students, who develop their understanding through the process of manipulating the sketch and describing what they observe. Encourage them to write clear and detailed explanations (and to use complete sentences) when they answer the questions; the extra time it takes them to do so is well spent.

If there's time and you have a presentation computer with a projector, have different students use Sketchpad to demonstrate to the class their observations or the problems they made up. It's a big help to students if they can listen to, evaluate, and discuss the descriptions and conclusions of their classmates.

INVESTIGATE

Students may be unfamiliar with *model* as a transitive verb; consider reviewing with them the various uses of this word.

Q1 In their final positions, the second arrow starts from where the first arrow ends, and the answer (13) is at the end of the second arrow. Encourage students to be detailed and specific in their answer to this question.

Q2 Answers will vary but should include only positive numbers.

Q3 Each lower arrow is exactly the same size and direction as the corresponding upper arrow.

Q4 The sum of $-6 + (-3)$ is -9.

Q5 Answers will vary but should include only negative numbers.

Q6 Whether adding two negative or two positive numbers, both arrows go the same way, taking the sum farther away from the center of the number line (farther away from zero). The difference is that the arrows go to the right when the numbers are positive but go to the left when they're negative.

Q7 When you add two negative numbers, you cannot get a positive sum. Both numbers take the sum in the negative direction from zero, so the sum must be negative.

Q8 As students model various problems, walk around the room and observe them to make sure they can model any problem they are given.

$$7 + (-4) = 3 \qquad -4 + 7 = 3$$
$$-6 + 2 = -4 \qquad 2 + (-6) = -4$$
$$-3 + 7 = 4 \qquad 3 + (-7) = -4$$
$$2 + (-5) = -3 \qquad -2 + 5 = 3$$

Q9 When you add a positive and a negative integer, the number that has the larger absolute value tells you whether the answer will be positive or negative. In other words, the sign of the result is the same as the sign of the longer arrow.

EXPLORE MORE

Q10 Each student will model different problems. In every case, the two numbers must be opposites, so that their arrows are the same length but point in opposite directions.

Q11 The order does not matter when you add two numbers. The arrows determine how far you go and in which direction, and it doesn't matter if you follow the first arrow and then the second, or if you follow the second arrow and then the first.

WHOLE-CLASS PRESENTATION

Start the whole-class presentation by animating the addition of two positive integers (Q1–Q3 of the activity). Open the sketch **Adding Integers Present.gsp** and press the step-by-step buttons one at a time, pausing between animations. Ask students to describe what they see as the animation progresses, and be sure to get observations from several different students. Press the *Reset* button, change the problem by dragging both circles (while leaving the numbers positive), and press the step-by-step buttons again.

Next animate the addition of two negative numbers (Q4–Q7 of the activity). Press *Reset,* drag the numbers so they are both negative, and ask students to predict what will happen now. Use the step-by-step buttons to test their conjectures. Without resetting, ask questions Q6 and Q7, and experiment by dragging to change the values of the numbers.

When students are satisfied with the results of adding two negative numbers, animate the addition of numbers with different signs. Reset again and drag the circles so one of the numbers is positive and one is negative. Ask students to predict how the arrows will behave. (Try to get students to concentrate on the behavior of the model rather than on the numeric answer.) Use the buttons again to show the behavior. Model several more problems (such as those in Q8) involving a positive and a negative number.

Finish the class discussion using Q9, Q10, and Q11. When students propose an answer to one of these questions, have them manipulate the sketch to show why their answer makes sense.

Adding Integers

In this activity you'll add integers using an animated Sketchpad model.

INVESTIGATE

1. Open **Adding Integers.gsp.** This sketch models the addition problem 8 + 5.

2. Press the *Present All* button to see the model in action.

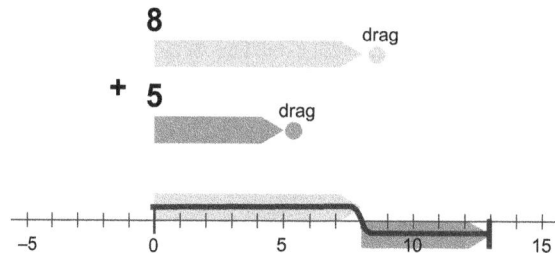

8

drag

+ 5

drag

−5   0   5   10   15

Q1 How does the final position of the arrows show the answer for 8 + 5?

3. Press the *Reset* button, and then drag the circles to model 2 + 6.

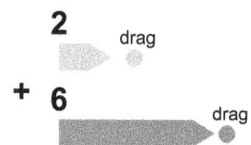

2 drag

+ 6 drag

4. This time, show the animation step by step: Press the *Show Steps* button, and then press each numbered button in order.

For each problem, press the buttons to show the result.

Q2 Drag the circles and press the buttons to model two other addition problems using only positive integers. Record each problem and the result.

Q3 How do the two upper arrows in the sketch relate to the two lower arrows?

Q4 Model −6 + (−3). What's the sum?

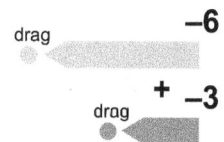

Q5 Model two more addition problems using negative integers. Record each problem and its result.

drag **−6**

+ −3 drag

Q6 How is adding two negative numbers similar to adding two positive numbers? How is it different?

Q7 Can you add two negative numbers and get a positive sum? Explain.

Q8 Model the following eight problems. Record each problem and its answer.

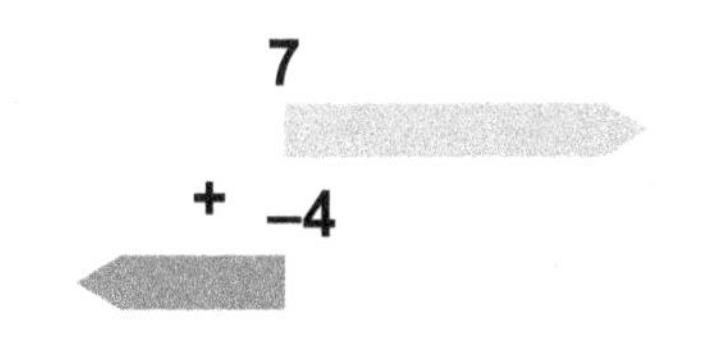 7 **+** −4	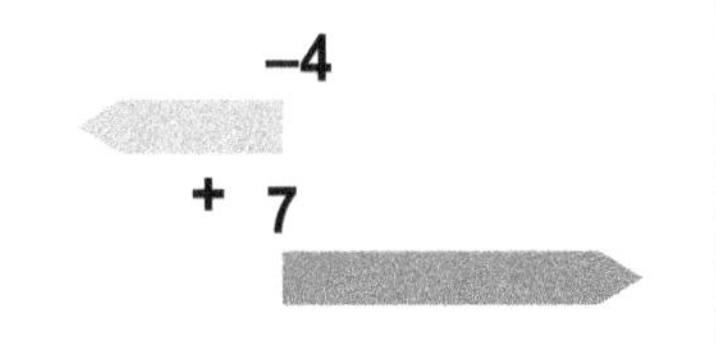 −4 **+** 7
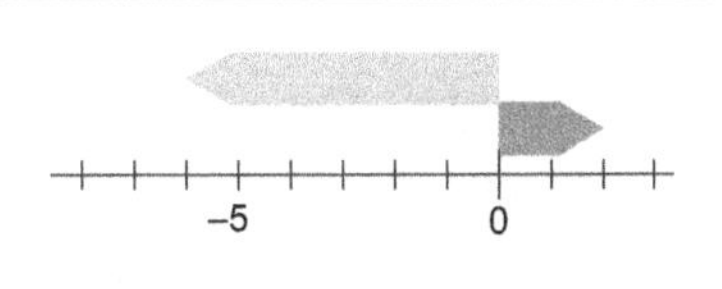	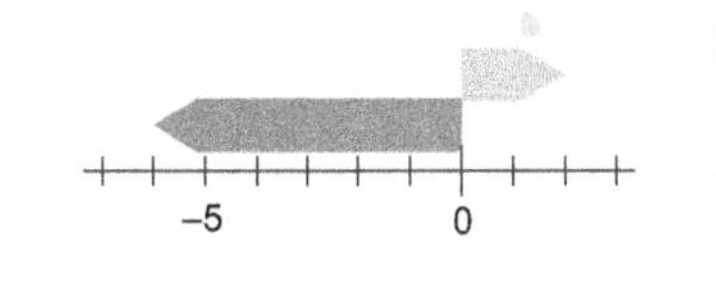
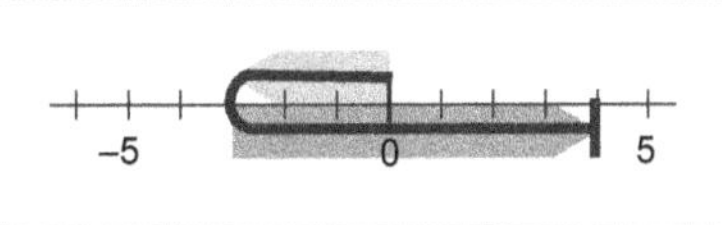	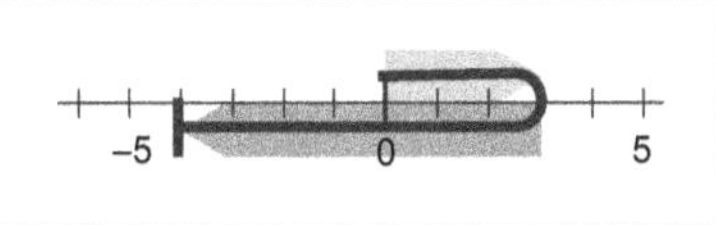
$2 + (-5)$	$-2 + 5$

Q9 When you add a positive and a negative integer, how can you look at the numbers and tell whether the answer will be positive or negative?

EXPLORE MORE

Q10 Model four problems for which the sum is zero. Make the first number positive in two problems and negative in two problems. Write down the problems you used. What must be true about two numbers if their sum is zero?

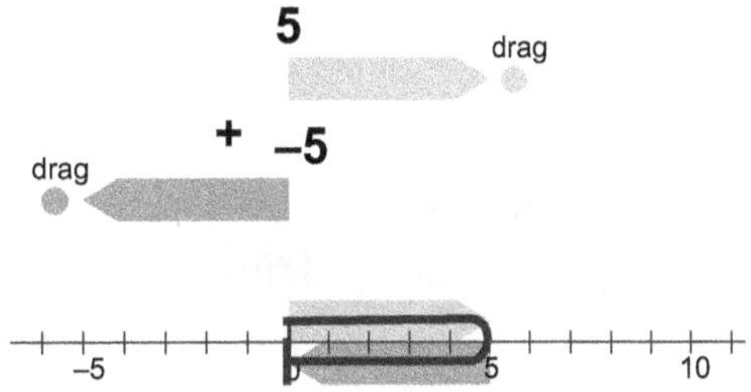

5   drag

+ −5

drag

Q11 When you add two numbers, does the order matter? In other words, is $-3 + 5$ the same as $5 + (-3)$? Using the sketch, explain why your answer makes sense.

Subtracting Integers

Use this activity as an introduction to integer subtraction for pre-algebra students, as a start-of-the-year refresher for Algebra 1 students, or as a supplemental activity for any student having difficulty with the topic. It's important for students to have a mental image of operations on integers. Even strong students who rely on verbal rules make careless mistakes that could be avoided by having an internalized picture.

The picture of subtraction presented here is a geometric model in which each number is represented by a vector. (The activity calls them *arrows* because students may not be familiar with the term *vector*.) Vectors incorporate both magnitude and direction (representing the absolute value and the sign of the integer), so practice with this model helps students understand how the signs of the operands come into play.

The questions are critical in encouraging students to internalize the model presented in this activity. Make sure students write clear and detailed explanations (and use complete sentences) when they answer the questions; the extra time it takes them to do so is time well spent.

If there's time and you have a presentation computer with a projector, have different students use Sketchpad to demonstrate to the class their observations or the problems they made up. It's a big help to students if they can listen to, evaluate, and discuss the descriptions and conclusions of their classmates.

INVESTIGATE

These notes sometimes use the terms *minuend* (first number) and *subtrahend* (second number), but these terms are not used in the student material. If you do use them with students, be sure to explain them carefully.

The concept of *additive inverse* is not named, but it plays a prominent role in the animation. You should discuss with the class why the second number must be flipped, even if you don't give a name to that operation.

Q1 During the animation, the arrow for 5 flips from the right to the left. This shows which way the second arrow must go in order to subtract it from the first.

Q2 In their final positions, the flipped second arrow starts from where the first arrow ends, and the answer (3) is at the end of the second arrow. Encourage students to be detailed and specific in their answer to this question.

Q3 Answers will vary. Students should describe the arrow flipping from right to left; encourage them to explain in their own words why it needs to flip in order to do subtraction.

Q4 Answers will vary but should include only problems in which a positive minuend is smaller than a positive subtrahend.

Q5 If both numbers are positive, the result will be positive if the first number (minuend) is larger, and negative if the second number (subtrahend) is larger.

Q6 Some students will record direct observations, and others will interpret those observations. Typical answers will be similar to the following.

Observation: In this problem, $4 - (-3)$, the second arrow starts out pointing to the left, so when it flips it turns around and points to the right.

Interpretation: The second number starts out negative, so when it flips it becomes positive.

Q7 The problems students create will vary. Because the first number is positive and the second negative, the models have in common that, after flipping, both arrows point to the right, and the result must be positive.

Q8 Problems will vary. Because the first number is negative and the second positive, after flipping, both arrows point to the left, and the result is negative.

Q9 As students model various problems, walk around the room and observe them to make sure they can model any problem they are given.

$$7 - (-4) = 11 \qquad\qquad -4 - 7 = -11$$
$$-6 - (-2) = -4 \qquad\qquad -3 - (-6) = 3$$
$$-3 - 8 = -11 \qquad\qquad -3 - (-8) = 5$$
$$2 - (-7) = 9 \qquad\qquad -2 - 7 = -9$$

Q10 Written as addition problems, these problems become

$$7 + 4 = 11 \qquad\qquad -4 + (-7) = -11$$
$$-6 + 2 = -4 \qquad\qquad -3 + 6 = 3$$
$$-3 + (-8) = -11 \qquad\qquad -3 + 8 = 5$$
$$2 + 7 = 9 \qquad\qquad -2 + (-7) = -9$$

In each case, to subtract you can change the sign of the second number and add them. This is similar to the way the second arrow flips before the animation shows the answer.

EXPLORE MORE

Q11 For a subtraction problem to have an answer of zero, the two numbers being subtracted must be the same.

Q12 To make the difference the same as the first number, the second number must be zero.

Q13 To make the difference the same as the second number, the first number must be twice as big as the second. For instance, $6 - 3 = 3$, and $-8 - (-4) = -4$.

Q14 The order does matter when you subtract numbers, because only the second arrow is flipped. More sophisticated students will observe that the order matters only if the second number is nonzero, because flipping zero has no effect.

WHOLE-CLASS PRESENTATION

Start the whole-class presentation by animating the subtraction of two positive integers (Q1–Q5 of the activity). Open the sketch **Subtracting Integers Present.gsp** and press the step-by-step buttons one at a time, pausing between animations. Ask students to describe what they see as the animation progresses, and be sure to get observations from several different students. Press the *Reset* button, change the problem by dragging both circles (while leaving the numbers positive), and press the step-by-step buttons again. Pay special attention to Q3 and Q5.

Next animate subtraction problems in which the first number is positive and the second number is negative (Q6–Q7 of the activity). Press *Reset*, make the first number positive and the second negative, and ask students to predict what will happen now. Test their conjectures using the step-by-step buttons. Repeat for several more problems.

Animate subtraction problems like those in Q8 and Q9, and record the answers for each of the problems in Q9. Ask students what patterns they see, and how they could predict the answer from the two numbers being subtracted.

For Q10, ask students to make an addition problem for each of the problems from Q9, and test their addition problems using page 2 of the sketch. Switching back and forth between page 1 and page 2 will reinforce for students the idea of using addition to rewrite a subtraction problem.

Continue the class discussion with as many of the Explore More questions (Q11–Q14) as are appropriate for the class and the available time.

Finish by having students summarize in their own words the relationship between subtraction and addition.

Subtracting Integers

In this activity you'll subtract integers using an animated Sketchpad model.

INVESTIGATE

1. Open **Subtracting Integers.gsp.** The sketch models the subtraction problem $8 - 5$.

2. Press the *Present All* button to see the model in action.

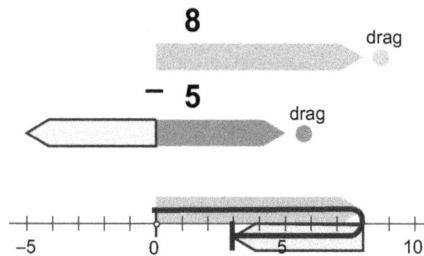

8
drag

$-$ **5**
drag

+—+—+—+—+—+—+—+—+—+—+
−5   0   5   10

Q1 During the animation, what happens to the arrow for 5?

Q2 How does the final position of the bottom arrows show the answer for this subtraction problem?

3. Press the *Reset* button, and then drag the circles to model $2 - 6$.

2
drag

$-$ **6**
drag

4. This time, show the animation step by step: Press the *Show Steps* button, and then press each numbered button in order.

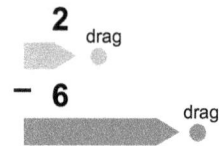

Q3 Describe in your own words what the *3. Make Inverse* step does.

For each problem, press the buttons to show the result.

Q4 Drag the circles to model two more subtraction problems that use positive integers but have a negative result. Record each problem and its result.

Q5 If both numbers in a subtraction problem are positive, how can you tell if the answer will be positive or negative?

Q6 Model $4 - (-3)$. What's different about the *3. Make Inverse* step this time?

4
drag

$-$ **−3**
drag

Q7 Model two more problems in which the first number is positive and the second number is negative. Record each problem. What do these models have in common?

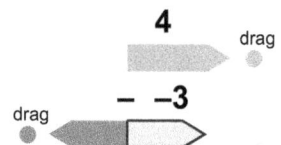

Q8 Model three problems in which the first number is negative and the second number is positive. Record each problem. What do these models have in common?

Q9 Model the following eight problems. Record each problem and its answer.

For instance,
7 − (−4) = 11,
so fill in the blank:
7 + ___ = 11.

Q10 For each subtraction problem above, write an addition problem that has the same first number and the same answer. What do you notice?

EXPLORE MORE

Q11 Model four subtraction problems for which the difference is zero. Make the first number positive in two problems and negative in two problems. Write down the problems you used. What must be true about two numbers if their difference is zero?

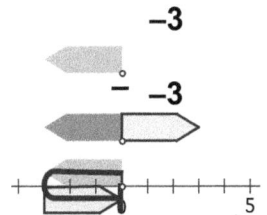

Q12 Model four subtraction problems in which the difference is the same as the first number. What must be true of these problems?

Q13 Model four subtraction problems in which the difference is the same as the second number. What must be true of these problems?

Q14 When you subtract two numbers, does the order matter? In other words, is −3 − (−5) the same as −5 − (−3)? Explain in terms of the model why your answer makes sense.

Dividing Real Numbers

ACTIVITY NOTES

MULTIPLICATION MACHINE

Q1 Marker *a* determines how many rectangles appear in the blue bar. Marker *b* determines the width of each blue rectangle.

Q2 When *a* shows exactly one yellow square, markers *b* and *c* move in unison.

Q3 You can calculate the length of the row of blue tiles by multiplying the number of tiles by the length of each tile. As an equation, $c = a \cdot b$.

DIVISION MACHINE

8. This is the simplest step of the activity, but in many ways the most important. Encourage students to observe closely and to think about what they observe.

Q4 When students press the button, the control for *b* goes below the line and the control for *c* goes above it, and students find that they can now control *c* rather than *b*. This is what makes it a division machine now: It calculates $b = c \div a$.

Geometrically, students are now controlling the size of the blue rectangles by dragging the end of the bar (representing the product) rather than by dragging the end of the first rectangle (the multiplier).

Q5 You would divide *c* by *a*. $b = c \div a$

Q6 $b = 5$ $(15 \div 3 = 5)$

$b = 3$ $(7.2 \div 2.4 = 3)$

Q7 Emphasize to students the importance of these four questions in developing and demonstrating their skills of translation and interpretation. Students must translate the questions, which are in mathematical terms, into the terms of the model and the behavior of the markers; must manipulate the model appropriately; and then must interpret the behavior they observe, expressing their conclusions mathematically.

a. When you drag *a* to exactly one, markers *b* and *c* are at the same value, no matter where you drag *c*. Dividing by one does not change the number.

b. No. When you drag *a* to be negative (for example, -2) and also drag *c* to be negative (for example, -6), the result is positive. As an equation, $-6 \div (-2) = 3$. Dividing by a negative number always

14

New York City Title I High School Activities with The Geometer's Sketchpad
© 2012 Key Curriculum Press

changes the sign, so dividing a negative by another negative gives a positive answer.

c. No. When you drag *a* between zero and one, the result is that *b* is larger than *c*. This means that dividing a positive number by a number between zero and one results in a larger number.

d. When you drag *a* toward zero while *b* is positive, the marker for *b* moves off the screen, showing a larger and larger result. When you get *a* to exactly zero, the machine breaks and the entire blue bar disappears. If you try the same thing while *c* is already zero, *b* also stays at zero as *a* gets close to zero, but when *a* is exactly zero, the machine breaks and *b* disappears.

Algebraically, if $c \div 0 = b$, then $c = 0 \cdot b$. If *c* is not zero, then there is no number *b* that will satisfy the second equation, because $0 \cdot b = 0$. On the other hand, if $c = 0$, then any real number will do for *b*, so there is no unique answer.

WHOLE-CLASS PRESENTATION

To present this activity to the whole class, start on page 2 of the sketch **Division Machine.gsp** and drag markers *a* and *b* to see how the machine performs multiplication. Ask students to explain the role of each variable in determining how the machine works. It's helpful to relate the behavior of this machine to the grouping model in the Multiple Models of Multiplication activity.

Once students are satisfied with the multiplying machine, press the *Multiply/Divide Toggle* button and observe the changes in the sketch. Switch this toggle several times to make sure students have seen the changes clearly, and then observe the behavior of the division machine's markers. As the class explores the parts of Q7, it may be helpful to switch back and forth several times between multiplication and division.

Dividing Real Numbers

When you first learned about division, your teacher probably began with a problem about distributing something fairly, such as "Divide 12 marbles among three children." Problems like this go only so far, because marbles and children come in positive integers. Division has to work for negative numbers and fractions.

MULTIPLICATION MACHINE

To understand division, you must first understand multiplication, so this activity starts with a multiplication machine.

1. Open **Division Machine.gsp.**

Q1 Experiment by dragging markers *a* and *b*. How can you control the number of rectangles in the blue bar? How can you control their width?

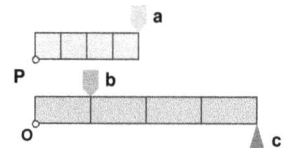

Q2 How do *b* and *c* behave when *a* is set to show exactly one yellow square?

To do real multiplication and division, you should have numbers, and for that you need a number line. You can follow steps 2–6 to make your own number lines, or you can go to page 2 of the sketch and skip to Q3 on the next page.

2. Press the *Show Number Line* button. Attach the blue bar to the number line by merging point *O* on the blue bar and point *O* on the number line.

> To merge the points, use the **Arrow** tool to select both of them and then choose **Edit | Merge Points.**

You'll need a separate number line for the yellow bar so that the bars don't overlap.

3. Construct a line through point *O*, perpendicular to the existing number line, by using the **Arrow** tool to select point *O* and the line and then choosing **Construct | Perpendicular Line.**

4. Use the **Number Line** custom tool to create the second number line. Press and hold the **Custom** tool icon, and choose **Number Line** from the menu that appears. Use the tool by clicking twice: first on the *Unit Distance* measurement and then on the perpendicular at the position where you want the new number line to appear.

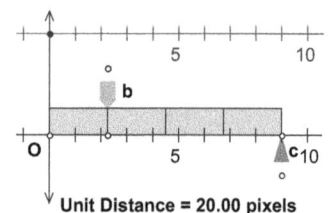

5. Choose the **Arrow** tool, and attach the yellow bar to the new number line by merging point *P* with the zero point of the number line.

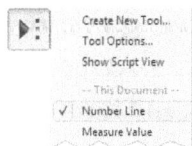

For GSP5

To hide points, use the **Arrow** tool to select them, and then choose **Display | Hide Points.**

6. To measure the position of each marker on the number line, use the **Measure Value** custom tool. Click each of the three points that appear above or below markers *a*, *b*, and *c*. Hide the three points.

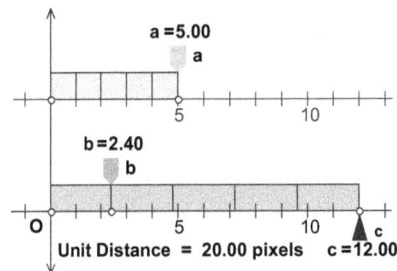

Q3 If you know the length of one blue tile and the number of tiles, you should be able to calculate the length of the row of blue tiles. How would you do this calculation? Write your answer as an equation using *a*, *b*, and *c*.

Choose **Number | Calculate.** Click the measurements to enter them in the calculation.

7. Use the measurements of *a* and *b* to calculate $a \cdot b$. Drag *a* and *b* to make sure that your calculation is correct.

DIVISION MACHINE

8. Press the *Multiply/Divide Toggle* button.

Q4 What changed when you pressed the button? Which markers can you control now? How is the behavior of the machine different?

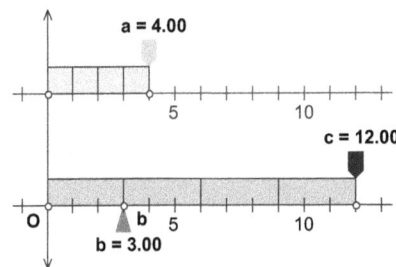

If you can't get the exact numbers you want, click the *Round* button for the value you're moving, and then try again.

Q5 By construction, $c = a \cdot b$. If you know the values of *a* and *c*, how would you use those numbers to calculate *b*?

Q6 Use the division machine to calculate $15 \div 3$. (To do this, drag *c* to 15 and drag *a* to 3.) What is *b*? Then use the same method to calculate $7.2 \div 2.4$.

Q7 Use your division machine to answer the following questions. For each question, tell how you dragged markers *a* and *c* to investigate, describe your observations, and explain your answers.

  a. What is the result when you divide a number by one?

  b. If you divide by a negative number, is the answer always negative?

  c. If you divide a positive number, is the answer always a smaller number?

  d. What happens when you try to divide by zero?

Ratio and Proportion

INVESTIGATE

Q1 The ratio shows a fraction representing the height of the rectangle divided by its width.

Q2 As you drag point *Adjust,* the yellow rectangle changes size and shape. It remains the same shape as the blue rectangle, and the same relative size.

Q3 As you drag point *Multiplier,* the size (but not the shape) of the yellow rectangle changes. When the multiplier is 1.000, the two rectangles are the same size.

Q4 You can drag point *Multiplier* until its value is 0.500. You can also look at the height and width measurements.

Q5 To make the yellow rectangle twice as big, drag point *Multiplier* until its value is 2.000. The height of the yellow rectangle is 20.00, and the width is 30.00.

Q6 To make the yellow rectangle's height 6 when the blue one's height is 8, you must use a multiplier of 0.750. The resulting width of the yellow rectangle is 7.50.

Q7 $w = 15$, $h = 5$, $x = 0.34$, and $m = 17.78$.

EXPLORE MORE

Q8 Answers will vary. One easy way to detect incorrect proportions is to set the multiplier to 1.00 and inspect the resulting fractions. Some of the incorrect ones (such as the one on the left below) can easily be identified.

$$\frac{10.00}{15.00} = \frac{15.00}{10.00} \qquad \frac{10.00}{15.00} = \frac{10.00}{15.00}$$

This method allows students to eliminate choices *b, c, e,* and *g.*

Another method is to adjust the multiplier so that the two ratios that make up a particular proportion have equal numerators, and then inspect the denominators. To test proportion *d,* you could manipulate the multiplier to be 1.50, causing *d* to appear as shown here. This proportion is obviously false because the two numerators match but the denominators don't.

$$\frac{15.00}{22.50} = \frac{15.00}{10.00}$$

The correct proportions are *a, f,* and *h.*

Q9 Here are the correct proportions expressed numerically and symbolically:

$$(a) \quad \frac{10}{15} = \frac{20}{30} \qquad \frac{h_1}{w_1} = \frac{h_2}{w_2}$$

$$(f) \quad \frac{20}{30} = \frac{10}{15} \qquad \frac{h_2}{w_2} = \frac{h_1}{w_1}$$

$$(h) \quad \frac{30}{20} = \frac{15}{10} \qquad \frac{w_2}{h_2} = \frac{w_1}{h_1}$$

You can use the square on page 4 to check the results on page 2. If you press the *Show Captions* button, you can edit the captions to change them to the corresponding numbers. The numbers will be displayed at the corners of the square, making it easy to check the proportions by transforming the square.

Q10 Here are the eight ratios that can be arranged in pairs:

$$\frac{h_1}{w_1} = \frac{h_2}{w_2} \qquad \frac{h_1}{h_2} = \frac{w_1}{w_2} \qquad \frac{w_1}{h_1} = \frac{w_2}{h_2} \qquad \frac{h_2}{h_1} = \frac{w_2}{w_1}$$

Here are the four ratios that cannot be arranged in pairs:

$$\frac{h_1}{w_2} \qquad \frac{w_1}{h_2} \qquad \frac{w_2}{h_1} \qquad \frac{h_2}{w_1}$$

WHOLE-CLASS PRESENTATION

In this presentation students will observe details of a model of a proportion using similar rectangles, manipulate the proportion by manipulating the model, use the model to solve various proportion problems, and investigate how any proportion can be written as an equation in eight different (but symmetrical) ways.

Begin by exploring the ratio of side lengths in a rectangle.

1. Open the sketch **Proportion Present.gsp** and drag point *Adjust.*

Q1 Ask, "What does dragging the point do to the rectangle?" After several answers, ask what two things about the rectangle are changed by moving point *Adjust,* and get a student to summarize that the point changes both the size and the shape of the rectangle.

2. Press the button labeled *Show Shape Buttons.* Press the shape buttons in turn to illustrate several different shapes.

3. Put the rectangle back into its original shape.

4. Press the *Show Ratio* button.

Q2 Ask, "What does the ratio represent?" Don't expect or impose any specific answers, but explore this question by changing the rectangle's shape.

Q3 Use either point *Adjust* or the shape buttons to make the rectangle tall and thin, and ask, "Is the ratio now a large number or a small one?"

Q4 Make the rectangle short and squat, and ask whether this ratio is a large number or a small one.

Q5 Ask, "What do you think will happen to the ratio if I press the *Square* button?" Press the button to test their conjectures.

Now add a second rectangle and look at its ratio.

5. Press the *Show Yellow Rectangle* button and drag point *Adjust*.

Q6 Ask, "What relationships do you see between the two rectangles as I drag *Adjust*?" Try to elicit student observations about both the relative shapes and the relative sizes.

6. Press the *Show Multiplier* button and drag point *Multiplier*.

Q7 Ask, "How does point *Multiplier* affect the relationship between the rectangles? How does it affect the relative shapes? How does it affect the relative sizes? What do you notice when the multiplier is 1.00? When it's 2.00?"

7. Set the multiplier back to 0.5. Press the *Show Second Ratio* button.

Q8 There's an equal sign between the ratios; ask, "Are the two ratios really equal?" Point out that the fact that they are equal means that the equation shown is a *proportion*.

8. Change the shape of the rectangles by dragging point *Adjust* or by pressing the shape buttons.

Q9 Ask, "Are the ratios still equal?" Ask, "How do you think the value of the multiplier is related to the numbers making up the two ratios?" Look at differently shaped rectangles to confirm students' conjectures.

Next use the rectangles to solve some proportion problems.

9. Go to page 2 and press the *Problem 1* button to display a proportion problem.

10. Have a student manipulate point *Adjust* so that the left side of the bottom proportion matches the left side of the problem. (When the point is close to the correct position, you can use the *Ratio of Integers* button to move it to the exact integer position.)

11. Have the student drag point *Multiplier* until the height of the yellow rectangle matches the number on the right side of the problem.

Q10 Ask, "Can you find the missing number in the problem by looking at the bottom proportion? What is the missing number?"

12. Problems 2, 3, and 4 on this page present additional challenges. To solve problems 3 and 4, show the size controls and press the smaller or larger buttons to make the scale of the rectangles appropriate for the numbers in the problem.

Explore how a particular proportion can be written in several different ways.

13. Go to page 3 and use the Calculator to compute various ratios. Make sure that students see that there are a number of different proportions that they can write using the same set of numbers.

14. The square on page 4 presents an animation of the symmetry involved in the eight different ways of expressing the same relationship. Experiment with it to see the various ways of expressing the same proportion.

Ratio and Proportion

In this activity you will use a visual model of ratios and proportions in similar rectangles to solve problems involving proportions.

INVESTIGATE

1. Open **Proportion.gsp.** Drag point *Adjust* to see what it does.

Adjust

2. A proportion requires two ratios. Show the first one by pressing the *Show Ratio* button. Then drag point *Adjust* again.

Q1 What does this ratio show?

3. The second ratio will come from a second rectangle. Press the *Show Yellow Rectangle* button, and then drag *Adjust* again.

Q2 What happens to the yellow rectangle as you drag *Adjust*? What determines the shape and size of the yellow rectangle?

4. Press the *Show Multiplier* button and then drag point *Multiplier.*

Q3 What happens to the yellow rectangle?

Q4 How can you make each side of the yellow rectangle half as big as the corresponding side of the blue one?

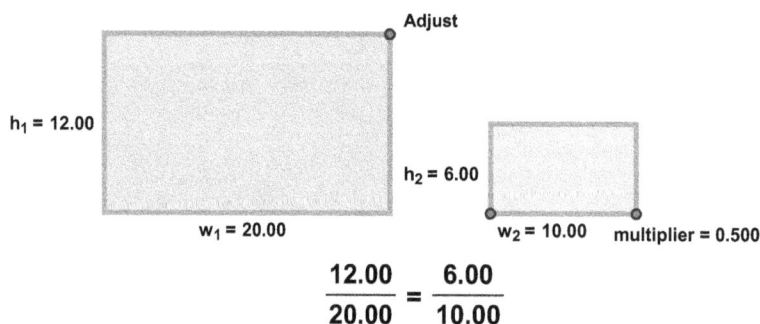

Adjust

$h_1 = 12.00$

$h_2 = 6.00$

$w_1 = 20.00$ $w_2 = 10.00$ multiplier = 0.500

$$\frac{12.00}{20.00} = \frac{6.00}{10.00}$$

5. To complete the proportion, press the *Show Second Ratio* button, and again drag points *Adjust* and *Multiplier.* Observe the effects on the proportion when you drag these two points.

The *Ratio of Integers* button can help you get the numbers exact.

6. Adjust the blue rectangle so its height-to-width ratio is 10.00/15.00. Press the *Set Multiplier to 1* button, and then make the yellow rectangle twice as big as the blue one horizontally and vertically.

Q5 What did you do to make the sides of the yellow rectangle twice as big as the sides of the blue one? What are the height and width of the yellow rectangle?

7. Adjust the blue rectangle so that h_1/w_1 is 8.00/10.00, and adjust the multiplier so that the yellow rectangle's height (h_2) is 6.00.

Q6 What multiplier did you use to do this? What is the resulting width (w_2) of the yellow rectangle?

Q7 In step 7, you actually solved the proportion $\frac{8.00}{10.00} = \frac{6.00}{w_2}$. Use the rectangles to solve the following proportions:

Show the size controls if you need to change the size of the rectangles.

$$\frac{8}{10} = \frac{12}{w} \qquad \frac{2.00}{3.00} = \frac{h}{7.50} \qquad \frac{x}{0.30} = \frac{1.70}{1.50} \qquad \frac{32}{m} = \frac{81}{45}$$

EXPLORE MORE

Go to page 2. This page lists eight different proportions using the height and width of the two rectangles. Some of the proportions are correct mathematically, but some are wrong. Determine which ratios are correct by manipulating the rectangles.

You can identify some incorrect proportions by making the rectangles square, and you can identify others by setting the multiplier to 1.

Q8 How can you tell which proportions are correct by manipulating the rectangles?

Q9 For each correct proportion, write down its letter, then the proportion as it appears using numbers, and finally the proportion using the symbols h_1, w_1, h_2, and w_2 in place of the numbers. For instance, proportion (a) is $\frac{10}{15} = \frac{20}{30}$, so you would write

$$(a) \quad \frac{10}{15} = \frac{20}{30} \qquad \frac{h_1}{w_1} = \frac{h_2}{w_2}$$

Choose **Number** | **Calculate** to use Sketchpad's Calculator. Click a measurement in the sketch to enter its value into the Calculator.

Q10 On page 3, calculate each of the 12 possible ratios involving h_1, w_1, h_2, and w_2. Arrange the ratios in pairs that have equal values. Write a proportion for each such pair. Which ratios do not belong to such pairs?

Go to page 4. This page shows the four variables from the rectangles at the four corners of a square. By reflecting the square across one of its axes of symmetry or by rotating by a multiple of 90°, you can generate all possible correct proportions. Use the square to check your answers for the proportions from pages 2 and 3.

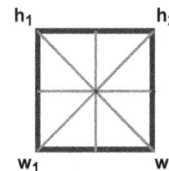

The Product of Two Binomials

It may take the students a few minutes to get the hang of placing new tiles. When using the custom tools to create tiles, it is very useful to click on an existing point in the sketch (as opposed to clicking in blank space or on some other type of object). This serves to anchor the new tile. The point you want to click on will be highlighted when the tool is positioned properly; when you see the point highlighted, it's time to click. If you make a mistake, just choose **Edit | Undo.** Improperly constructed tiles will not stay in alignment when you adjust the sliders.

It's important that students drag the sliders for x and y periodically—this is the big advantage of using Sketchpad algebra tiles, after all. First, dragging tests whether they've used the custom tools properly. But more importantly, it reinforces the fact that x and y are variables and that the relationships discovered work no matter what their values.

INVESTIGATE

2. When students use a custom tool to construct a new tile, the tile is anchored where they click, but its direction is not yet determined as horizontal or vertical. To orient the tile vertically or horizontally, students must use the **Arrow** tool to drag the black point on the arc near the upper-left corner of the tile.

   Some students may prefer to orient each tile as soon as they construct it. Others may prefer to attach several tiles at various angles and then orient them all later.

Q1 The term x^2 refers to the big square in the upper left of the rectangle you built inside the frame. It's the product of the x from $(x + 3)$ and the x from $(x + 5)$.

   The term $8x$ refers to all of the non-square tiles: $5x$ from the lower left and $3x$ from the upper right. The $5x$ is the product of the x from $(x + 3)$ and the 5 from $(x + 5)$. The $3x$ is the product of the x from $(x + 5)$ and the 3 from $(x + 3)$.

   The number 15 refers to the 15 unit squares at the lower right of the rectangle. It's the product of the 3 from $(x + 3)$ and the 5 from $(x + 5)$.

Q2 $y^2 + 5y + 4$

Q3 a. $x^2 + 5x + 6$

b. $2y^2 + 7y + 3$

c. $x^2 + xy + 2x + 2y$

d. $x^2 + 4x + 4$

e. $2x^2 + 5xy + 2y^2$

f. $6y^2 + 8y + 2$

EXPLORE MORE

Q4 Possible answers: Use an "opposite" color to represent negatives, or use dashed lines or shading. Students will come to appreciate the difficulty of representing negatives (particularly negative areas).

The Product of Two Binomials

Mono-, bi-, and *tri-* are prefixes from the Greek words for one, two, and three, respectively.

The expression $x + 3$ is called a *binomial* because it consists of two *monomial* terms: x and 3. The expression $(x + 3)(x + 5)$ is the product of two binomials, $x + 3$ and $x + 5$. In this activity you'll use Sketchpad algebra tiles to model expressions equivalent to the products of binomials. The process you'll learn, called *expanding*, is used for writing expressions in different forms and for demonstrating the equivalence of algebraic expressions.

INVESTIGATE

1. Open **Binomial Product.gsp.**

You'll see the factored expression $(x + 3)(x + 5)$ modeled with algebra tiles. The blue tiles represent x and the yellow tiles represent one unit. Notice that one binomial factor is modeled as a vertical train and the other is modeled as a horizontal train.

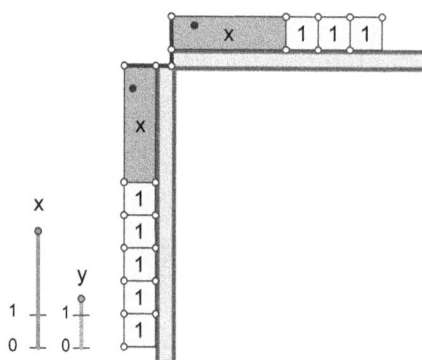

Press and hold the **Custom** tool icon to see the available tools. Choose the tool you want to use; then click a point in the sketch to construct that tile. You can drag the black point later to rotate the shape to the orientation you want.

2. Use the custom tools that come with the sketch to tile the rectangle, using tiles whose dimensions match the horizontal and vertical trains.

When you're done, adjust the x slider and see if your rectangle holds together.

The expression you modeled is $x^2 + 5x + 3x + 15$. You can combine the like terms to get $x^2 + 8x + 15$. This *trinomial* (an expression with three monomial terms) is called the *expanded form* of $(x + 3)(x + 5)$.

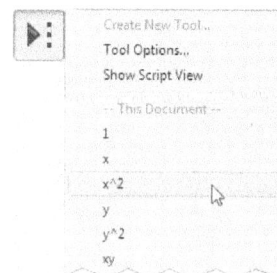

Q1 Explain how each of the terms in the trinomial $x^2 + 8x + 15$ is related to the product of the binomials.

3. Press *Hide/Show Values* to show the values of x and y.

Choose **Number** | **Calculate** to open the Calculator. Click the *x* measurement in the sketch, and type from your keyboard to build the expressions.

4. Use Sketchpad's Calculator and the measurement for *x* to calculate the values of the expressions $(x + 3)(x + 5)$ and $x^2 + 8x + 15$ for the current value of *x*.

5. Change the length of the *x* slider to confirm that the expressions remain equivalent for different values of *x*.

6. Go to page 2. Use the **y** and **1** custom tools to build trains representing the product $(y + 1)(y + 4)$. Attach each tile to a white corner of the frame or the previous tile.

To orient the tiles horizontally or vertically, drag the black arc points with the **Arrow** tool.

7. Orient the tiles horizontally or vertically, depending on where they are used.

8. Use the various custom tools to add tiles within the frame. Drag the *y* slider to make sure your tiling correctly represents this area.

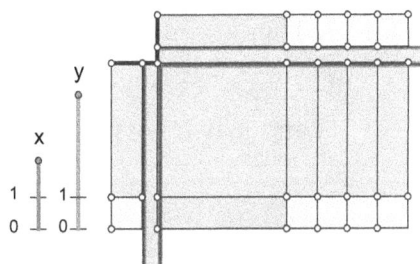

Q2 Write the expanded expression represented by the rectangle. Combine like terms.

9. Press *Hide/Show Values* to show the values of *x* and *y*. Then use Sketchpad's Calculator and the measurement for *y* to calculate the values of the expression $(y + 1)(y + 4)$ and the expanded expression you just found. Adjust the *y* slider to confirm that these expressions are always equivalent.

Q3 Build and expand the following expressions on the remaining pages of the sketch. Draw the models on paper. Write each expression both as a product of binomials and as a trinomial.

    a. $(x + 2)(x + 3)$ b. $(2y + 1)(y + 3)$ c. $(x + y)(x + 2)$

    d. $(x + 2)(x + 2)$ e. $(2x + y)(x + 2y)$ f. $(3y + 1)(2y + 2)$

EXPLORE MORE

Q4 Experiment with ways the model can be altered to represent expressions with negatives. For example, how could you represent $(x + 2)(x - 3)$? Illustrate and explain any models you think of.

TILING THE SQUARE

Q1 The x and y sliders change the sides of the square.

Q2 The side length of this square is x. (Students can test this by dragging it next to the x slider.) The area is x^2.

Q3 The side length of this square is y. (Students can test this by dragging it next to the y slider.) The area is y^2. The sum of the areas of these two squares is $x^2 + y^2$.

Q4 Generally, this is not possible. There is extra space in the outlined square that cannot be covered by the two smaller tiles. Some students may discover a special case. It is, in fact, possible if x or y is equal to zero. They can model this case by changing the control sliders.

Q5 These tiles will fill the square no matter what lengths are used for x and y. Two possible solutions are shown below:

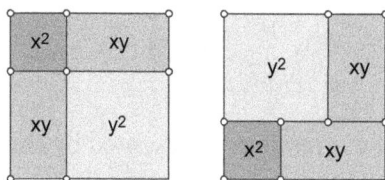

Q6 Below is one pattern that will work:

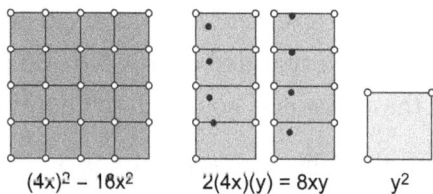

$(4x)^2 - 16x^2$ $\qquad$ $2(4x)(y) = 8xy$ $\qquad$ y^2

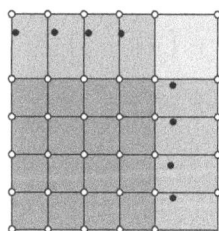

The equation is $(4x + y)^2 = 16x^2 + 8xy + y^2$.

BINOMIALS WITH SUBTRACTION

Q7 If the squares are covered as shown below, the only light color remaining will be a square with area $(x - y)^2$.

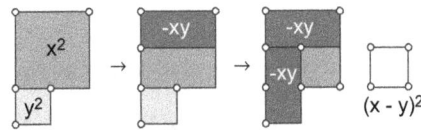

WHOLE-CLASS PRESENTATION

To present this activity to the whole class, use the Presenter Notes and the sketch **Squaring Binomials Present.gsp.**

Squaring Binomials

In this presentation students will see a visual representation of the squaring of a binomial, and will make a connection between the squares and rectangles on the screen and the various terms that make up the algebraic expression.

1. Open **Squaring Binomials Present.gsp.** The empty square represents the square of the binomial $x + y$.

2. Change the value of x by dragging the point at the end of the blue slider.

Q1 Ask, "What happens when the value of x changes? How does it affect the two colored squares? How does it affect the rectangles?" (Only one square, representing x^2, is affected, and only the x dimension of the rectangle changes.)

Be sure to leave the values of x and y significantly different.

3. Similarly, change the value of y and have students observe the effects.

4. Drag the colored shapes into the empty square and arrange them so they fill the empty square. To flip a rectangle to the correct angle, drag its black point.

Q2 Ask, "Geometrically, the shapes fit exactly. What does this mean algebraically?" Have students describe the connection between the shapes and the four terms on the right-hand side of $(x + y)^2 = x^2 + xy + xy + y^2$.

If you use the custom tools to attach the shapes, they will fill the square even when you change x and y.

5. Drag x or y to change the size of the empty square. The pieces are no longer in the correct positions. Press the *Tile* button to move the shapes so they fit again. Make sure students are convinced that the shapes will always fit.

6. Go to page 2, representing $(4x + y)^2$. Drag x and y to change the shapes. There are lots of tiles to move here, so use the *Tile* button right away.

Q3 Ask students to count the shapes and to write a formula based on the number of squares and rectangles they count: $(4x + y)^2 = 16x^2 + 8xy + y^2$.

The dark rectangles can only be used to cover up positive area.

7. Go to page 3. This page represents $(x - y)^2$. Drag x and y to show how the shapes change. Point out that the dark $-xy$ rectangles represent negative area.

8. Add the x^2 and y^2 tiles by placing them next to each other such that one vertex and one side coincide. Drag the $-xy$ rectangles so they cover as much of the positive area as possible.

Q4 Ask, "How much positive area is left after the subtraction?" Students may guess that the remaining positive area is equal to the $(x - y)^2$ square. Confirm this by dragging the $(x - y)^2$ square so it coincides with the remaining positive area.

9. Change the values of x and y, switching which of them is larger. Separate the tiles, and then use the *Tile* button to arrange them again.

Q5 Ask students to write the formula that the tiles illustrate:
$(x - y)^2 = x^2 + y^2 - xy - xy$.

Squaring Binomials

For GSP5

Ask an algebra teacher what the most common mistakes are among algebra students. This one is certain to be near the top of the list:

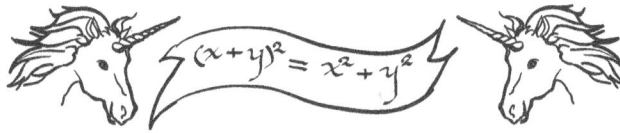

$$(x + y)^2 = x^2 + y^2$$

In general, this equation is not correct. There is no property that allows you to distribute exponents over addition this way. Yet even the best students continue to make this mistake years after they have learned better. The fallacy of this equation becomes clearer when you model it geometrically with the help of algebra tiles.

TILING THE SQUARE

1. Open **Squaring Binomials.gsp**.

Use the **Arrow** tool to drag the endpoints of the sliders.

Q1 Drag the x and y sliders. What effect do they have on the square?

The sides of the square are $x + y$, so its area must be $(x + y)^2$.

2. Press and hold the **Custom** tool icon. Choose the **x^2** tool from the menu that appears. Click in an open place on the screen to create the square.

Q2 In terms of x and y, how long is each side of the new square? What is its area?

3. Choose the **y^2** custom tool and use it to construct another square in open space.

Q3 In terms of x and y, how long is each side of this square? What is its area? What is the sum of the areas of the two squares you just constructed?

If the equation $(x + y)^2 = x^2 + y^2$ is correct, then the area of the outlined square (on the left side of the equation) must be equal to the sum of the areas of the two new squares you just constructed (on the right side of the equation).

Q4 Test whether the equation is true by dragging the two new squares into position so that they precisely cover the outlined square. Can you do this? Explain your answer.

Each *xy* has one black point. Drag this point to flip the rectangle.

4. In fact, $(x + y)^2 = x^2 + 2xy + y^2$, so you will need more tiles to fill the square. Choose the **xy** custom tool. Click twice on the screen. Choose the **Arrow** tool again.

Q5 Is it possible to fill the large square with the tiles you have created? Move the tiles into place to fill it. Draw a diagram showing how you did this.

What if you square a more complicated binomial? Consider $(3x + 2y)^2$. You can see four parts of the square separated in the image on the right. By counting shapes, you can see that $(3x + 2y)^2 = 9x^2 + 12xy + 4y^2$.

Q6 Now try it with a different binomial. Go to page 2. There is a square representing $(4x + y)^2$. Use the custom tools to form four groups of tiles, as shown in the preceding example. Then drag the tiles into the large square to fill it. Draw a diagram of the results and write the equation they represent.

BINOMIALS WITH SUBTRACTION

What if the binomial has a subtraction sign separating the two terms? In that case, the middle term of the expansion is negative:

$$(x - y)^2 = x^2 - 2xy + y^2$$

In order to show this geometrically, you need a way to show negative area.

The dark rectangles represent negative area.

5. Go to page 3 of the document. There are light-colored squares and two dark rectangles.

To add area, put two shapes next to each other. To subtract area, cover up a positive area with a negative area.

6. Add x^2 and y^2 by putting their squares next to each other. Then subtract $2xy$ by covering as much of the positive area as you can with the $-xy$ rectangles.

Q7 How much positive area is left? How does this remaining area compare with the $(x - y)^2$ square? Draw a diagram of your results from step 6.

2

Solving and Graphing Functions and Inequalities (Algebra)

Solving Linear Equations by Balancing

The balance used in this activity was made using the tools described on the Algebalance page of the presentation sketch. You can use these tools to create your own balance sketches.

EXPLORE THE BALANCE

Q1 The positive objects (1, 5, and x) weigh the pan down, and the negative ones (-1, -5, and $-x$) pull it up. (The behavior of x and $-x$ depends on whether x is positive or negative. This is explored later in the activity, so there's no need to worry about it yet.)

Q2 Answers will vary.

Q3 Answers will vary. The equation should match the list of objects from Q2.

Q4 Students should press *Setup Q4* before each part so that they have enough objects to perform the operation.

  a. The pans remain balanced.

  b. The right pan goes up and the left pan goes down.

  c. The pans remain balanced.

  d. The left pan goes down and the right pan goes up.

  e. The pans remain balanced.

  f. The pans remain balanced.

  g. The left pan goes up and the right pan goes down (if students have not found and adjusted the x slider).

  h. The pans remain balanced.

Q5 The rules are described later in the activity. Encourage students to answer this question in their own words.

Rule 1: You can drag the same kind of object from storage to each pan. (This rule is illustrated by Q4 parts a and e. Some students may also list f, which involves removing the same object from both pans.)

Rule 2: If you have matching positive and negative objects on a pan, you can remove them to the storage area. (This rule is illustrated by Q4 parts c and h.)

SOLVE AN EQUATION

Q6 The balance on page 2 represents $2x - 1 = x + 5$.

Q7 Rule 1 allows you to drag identical objects to each pan.

Q8 Rule 2 allows you to eliminate pairs like x and $-x$. Students can use this rule twice, once on each pan.

Q9 The resulting equation is $x + (-1) = 5$. The sketch is limited in how it can display equations, and the appearance in the sketch is $x + -1 = 5$.

Q10 Students can remove a single combination of 1 and -1 from the left pan.

Q11 The resulting value of x is 6.

MORE EQUATIONS

Q12 Here are the steps, though students may change the order in some ways. Students may also combine similar steps such as b and c, or they may use step g to add three 1's to each pan.

  a. Add $-x$ to both pans: $2x + (-3) = 1x + (-1)$
  b. Remove x and $-x$ from the left: same equation
  c. Remove x and $-x$ from the right: same equation
  d. Add $-x$ to both pans: $x + (-3) = -1$
  e. Remove x and $-x$ from the left: same equation
  f. Remove x and $-x$ from the right: same equation
  g. Add 1 to both pans: $x + (-2) = 0$
  h. Remove 1 and -1 from the left: same equation
  i. Remove 1 and -1 from the right: same equation
  j. Add 1 to both pans: $x + (-1) = 1$
  k. Remove 1 and -1 from the left: same equation
  l. Remove 1 and -1 from the right: same equation
  m. Add 1 to both pans: $x = 2$
  n. Remove 1 and -1 from the left: same equation
  o. Remove 1 and -1 from the right: same equation

The last formula ($x = 2$) shows the value x must have to make the pans balance.

Q13 On page 4, the balance shows $2x + 5 = x + 2$.

New York City Title I High School Activities with The Geometer's Sketchpad
© 2012 Key Curriculum Press

Q14 To solve this equation, here are the steps:

    a. Add $-x$ to both sides: $x + 5 = 2$

    b. Add -1 to both sides: $x + 4 = 1$

    c. Add -1 to both sides: $x + 3 = 0$

    d. Add -1 to both sides: $x + 2 = -1$

    e. Add -1 to both sides: $x + 1 = -2$

    f. Add -1 to both sides: $x = -3$

Students may consolidate some of steps b–f by dragging more than a single -1 at a time.

Q15 A single x on the left pan pulls it up, even though it weighed it down in Q1. In Q1 (on page 1) x was positive, causing it to weigh the balance down. Here, x is negative, so it pulls the pan up. If you drag the x slider so that x is positive, this page will give the same result as page 1. If possible, have students try this by putting a single x on the left pan and then dragging the x slider to both positive and negative values.

Q16 The solution to $4x - 2 - x = x + 3 + x$ is $x = 5$.

EXPLORE MORE

Q17 When each pan has two identical columns, removing one column from each pan will keep the pans in balance. This is equivalent to dividing the number of objects on each pan by two.

Q18 a. $x = 3$ b. $x = -3$ c. $x = -4$

Q19 $-2x + 1 = -3x + 4$: $x = 3$

    $2x - 2 = 3x + 1$: $x = -3$

    $-x + 3 = -2x + 1$: $x = -2$

Q20 Students will make different equations. If possible, have some students show their equations and the method to use in solving them.

Q21 It may help students to study page 7 to see how the buttons on that page work.

CLASS DISCUSSION

Some fundamental principles arise during this activity.

Rule 1 is the addition property of equality. Have students explain this rule in their own words during the discussion. If they are already familiar with this property by name, encourage them to make the connection between the concrete action of dragging objects on the screen and the abstract formulation of the property. If they are not familiar with the property by name, this may be a good opportunity to have students come up with their own name for the property that is more descriptive.

Different students will justify Rule 2 differently. A class discussion will help them develop insights into the concepts of additive identity and the additive inverses, and how they can be used to solve equations. There's no need to name these concepts; the important point is for students to realize that the sum of a number and its additive inverse is zero, so that removing the combination from one side of an equation leaves the two sides equal.

A discussion of the balance on page 4, and of Q15 in particular, provides an opportunity for students to recognize that $-x$ is not necessarily negative, any more than x is necessarily positive. This important realization can prevent many errors and misunderstandings later.

Finally, students are told several times during the activity to try to get a single x on a pan all by itself. The activity doesn't try to explain this strategy, so a class discussion is a good opportunity for students to think about this strategy and understand why it's useful.

WHOLE-CLASS PRESENTATION

In this presentation students see the effects of placing both positive and negative objects on the left and right pans of a balance, and connect the visible objects with a simplified formula corresponding to the current state of the balance. This presentation helps students develop a mental model of equations and of the techniques used to solve them.

Use **Solve by Balancing Present.gsp** and the Presenter Notes to conduct a presentation for the entire class.

Introduce the presentation.

1. Open **Solve by Balancing Present.gsp** and show how the balance works. You can press the numbered action buttons to show how the balance responds to the objects, or you can just drag objects to and from the balance.

   As you press the buttons or drag the objects, explain how an equation is like a two-pan balance, with equality corresponding to the pans being in balance and inequality corresponding to one pan being heavier than another.

Illustrate two rules students can use to solve equations.

2. Use page 2 to demonstrate one of the rules that makes it possible to solve equations by balancing: the addition property of equality. Press the *Show Premises* button to show the premises and to reveal buttons that allow you to illustrate the rule.

3. Use page 3 to demonstrate a second rule: Opposites on the same side of the equation add up to 0, and you can remove them.

Use the rules to solve several equations.

4. Use page 4 to show that you can solve simple equations using these rules. Press each numbered button in turn to show the various steps in solving the equation. Discuss how the solution becomes obvious once the balance has been manipulated so that there's only a single x left, on a pan by itself.

5. Use page 5 to solve another simple equation. The value of x on this page is negative, so that an x object pulls the balance up instead of weighing it down, and a $-x$ object pulls it up. After you finish solving the equation, press the *Reset* button to clear the pans, and show what happens when you drag the x and $-x$ objects on and off the pans. Discuss this with your students, and encourage them to describe and explain this behavior in their own words.

If students are not ready to think of the operation on this page as multiplying by $\frac{1}{2}$, describe it as dividing by 2.

6. Use page 6 to suggest a third rule (the multiplication property of equality). In this example both pans have two identical columns of objects. By removing half of what's on each pan (removing half the weight on each pan), the pans remain balanced. Similarly, doubling the number of objects on each pan keeps them balanced.

7. Page 7 shows the solution of a more complicated equation using the balance.

8. Page 8 is blank. You can use it to set up and solve any of the problems from the student activity pages.

Finish with a class discussion. As part of the discussion, encourage students to ask questions about and to justify these techniques for isolating x on one side of the equation while keeping it balanced.

Solving Linear Equations by Balancing

To solve a complicated equation, you can make it simpler while keeping the two sides of the equation equal. In this activity you will use a Sketchpad balance to show an equation and solve it by using operations that keep the sides balanced.

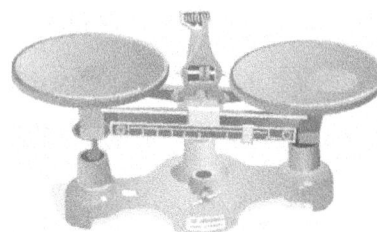

EXPLORE THE BALANCE

1. Open **Solve by Balancing.gsp.** Experiment by dragging objects from the storage area (on the left of the dividing bar) to the left or right balance pan.

Q1 Which objects weigh a pan down, and which ones pull it up?

Q2 Find a combination of weights and balloons different from the one shown below that balances the two pans. List the objects you put on each pan.

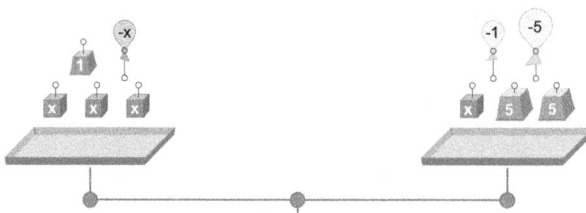

Q3 Write down the algebraic equation that corresponds to your arrangement of weights and balloons. Press the *Show Formula* button to check your answer.

When the pans are balanced, some operations disturb the balance and others do not.

Q4 Try each of these operations and write down how it affects the balance. Before each operation, press *Setup Q4* to make sure the pans are balanced and that each pan contains enough items to carry out that operation.

In your answer, state whether the operation makes the left pan heavier, makes the right pan heavier, or leaves the two pans in balance.

  a. Drag a 1 from the storage area onto each pan.

  b. Drag a 1 from the right pan to the left pan.

  c. Drag a 1 and a -1 together from the left pan to the right pan.

  d. Drag a 1 from the storage area onto the left pan and a -1 onto the right pan.

  e. Drag a -5 onto each pan.

  f. Remove an x from each pan by dragging to the storage area.

  g. From the storage area drag an x onto the left pan and a 5 onto the right pan.

  h. Drag an x and a $-x$ from the right pan to the storage area.

Q5 Write down at least two rules to describe things you can do that will keep the pans balanced. For each rule, write down which parts of Q4 illustrate the rule.

New York City Title I High School Activities with The Geometer's Sketchpad
© 2012 Key Curriculum Press

SOLVE AN EQUATION

Q6 Go to page 2. What equation does this balance represent? Press the *Show Formula* button to check your answer.

In the next few steps you will use these balancing rules:

Rule 1: You can drag the same kind of object from the storage area to each of the pans. (For example, you can drag a 5 to the left pan and a 5 to the right pan.)

Rule 2: If you have matching positive and negative objects on the same pan, you can remove them to the storage area. (For instance, if the left pan has both a 5 and a −5, you can remove them both.)

Make sure the pans stay in balance after each operation.

Q7 Drag a −x from the storage area onto each pan. Which rule allows you to do this?

Q8 Any time you find both an x and a −x on the same pan, you can remove them to the storage area. Which rule is this? How many such combinations can you find? Remove them now.

Q9 What is the resulting equation?

Q10 Drag a 1 from the storage area onto each pan. Any time you find both a 1 and a −1 on the same pan, you can remove them to the storage area. How many such combinations can you find and remove?

Q11 What is the resulting value of x? Press the *Show x* button to check your result.

MORE EQUATIONS

You have solved the equation when you have x by itself on one pan and only numbers on the other pan.

2. On page 3, use the rules about moving objects to eliminate as many objects as you can and to leave the last x all alone on its pan.

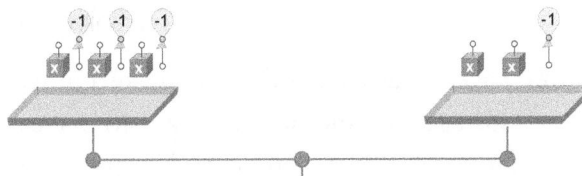

Q12 Write down each step that you follow, and write down the equation for the balance after each step.

Q13 On page 4, what equation does the balance show?

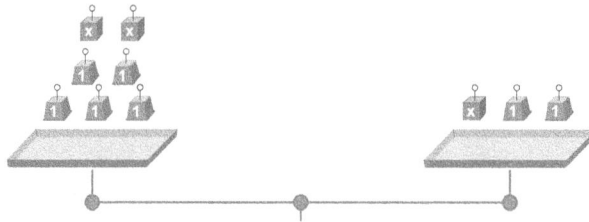

Q14 Use the two rules to add and remove objects until you get x on a pan by itself, with only numbers on the other pan. Each time you use Rule 1, write down what you did and the resulting equation. What is x?

Q15 Remove all the objects from the pans by pressing the *Reset* button, and then move a single x onto the left pan. Does this object weigh the pan down or pull it up? Is your answer the same as it was for Q1? If not, why not?

Q16 Go to page 5 and add objects to the pans to model the equation $4x - 2 - x = x + 3 + x$. Then solve the equation by following the two rules.

EXPLORE MORE

Q17 There's one more rule you can use with the algebra balance. Go to page 6 and notice that each pan has two identical piles of objects. Remove one pile from the left pan and one pile from the right pan. Do the pans stay in balance? What mathematical operation did you perform on each pan?

Q18 With the help of this third rule, solve these three equations:

 a. $2x - 1 = 5$ b. $4x + 3 = 2x - 3$ c. $4x + 5 = x - 7$

Q19 Page 7 contains several buttons to create different arrangements of objects. For each button, press the button and write down the equation that results. Then use the various rules you've learned to rearrange the objects and solve the equation, and write down the solution that you find.

Q20 Page 8 is a blank page. Show the value of x, adjust the slider, and arrange objects to make a balanced equation of your own choosing. Then hide the x value and slider, and challenge a friend to solve your equation.

Use Sketchpad's Help menu to find out how to make and use movement buttons.

Q21 Page 9 is also a blank page. Make a movement button to move objects out of the storage area and onto the pans to make an equation. Then make a movement button for each step in solving the equation. Use the movement buttons to demonstrate your problem to your classmates.

Solving Inequalities by Balancing

For GSP5 ACTIVITY NOTES

The balance used in this activity was made using the **Algebalance.gsp** sketch. You can use these tools to create your own balance sketches; instructions accompany the sketch.

EXPLORE IMBALANCE

Q1 **Rule 1:** Dragging the same type of object onto each pan does not change the balance.

Q2 **Rule 2:** Removing an object and its opposite (for instance, x and $-x$, or 1 and -1) from a pan does not change the balance.

Q3 The pans represent the inequality $x + 3 > 2x - 1$.

Q4 The solution is $4 > x$.

Q5 The original inequality is $3x - 2 < 2x - 3$. To solve it, follow these steps:

   a. Move two $-x$ objects onto each pan. (Rule 1)
   b. Remove zeros, resulting in $x - 2 < -3$. (Rule 2)
   c. Move two 1 objects onto each pan. (Rule 1)
   d. Remove zeros, resulting in $x < -1$. (Rule 2)

Q6 Removing exactly half of the objects on each pan cannot change the balance. Though the example involves multiplying by one-half (or dividing by two), Rule 3 actually states that you can multiply or divide both pans by any positive number. (Negative numbers work here for equality, but not for inequality.) Students will formulate Rule 3 in different ways, some more general and some more limited. This is a good question to pursue in a class discussion.

Q7 The inequality is $x + 3 > 3x - 3$. The solution is $x < 3$.

CLASS DISCUSSION

Discuss with the class the differences between using the balance to solve equations and using it to solve inequalities.

A very important difference to note is that it's possible to violate the rules without any visible effect, depending on the nature of the violation and the value of x. For instance, if the correct solution is $x < 3$, and a student makes a mistake and ends up with $x < 4$, she will not see any change in the state of the balance, even though she now has the wrong solution set. The state of the balance gives information for only a single value of *x,* not for the entire solution set.

A discussion of Rule 3 is also important, both because some students will formulate it in different ways in Q6 and because the rule is different for inequality than it is for equations. Students will understand Rule 3 better if they have done the activity Properties of Inequality.

Finally, review the value of the strategy in which students try to get a single *x* on a pan all by itself. Understanding the value of this arrangement of the balance will help students in solving equations and inequalities symbolically.

WHOLE-CLASS PRESENTATION

Use **Inequalities by Balancing.gsp** to conduct a presentation for the entire class. Follow the steps of the student activity, and involve the class in the process of answering the questions in the activity.

Solving Inequalities by Balancing

Just as you solve equations by keeping the two sides of the equation balanced, you can solve inequalities by keeping the two sides of the inequality unbalanced.

EXPLORE IMBALANCE

1. Open **Inequalities by Balancing.gsp.**

On page 1 the pans are balanced. Use this page to review the rules for what you can do without affecting the balance.

Q1 Drag a 1 from the storage area to each pan. Then drag a $-x$ to each pan. Do these steps disturb the balance of the pans? This is Rule 1. Write it down.

Q2 Drag an x and a -1 to each pan. Remove the x and $-x$ from the left pan. Remove the 1 and -1 from the right pan. When you remove two opposite objects from a pan, does it disturb the balance? This is Rule 2. Write it down.

Q3 On page 2 the pans are not balanced. Write down the inequality the pans represent.

As long as you follow the rules, you will not disturb the state of the pans.

Q4 Move objects on and off the pans (always following the two rules) until you get a single x all by itself on one pan and only numbers on the other pan. Write down this inequality.

Q5 Page 3 has a different arrangement, but the pans are still unbalanced. Use the two rules again to solve this inequality. Write down the original inequality, the steps you use, and the final inequality (when a single x is left on a pan by itself).

Rule 3 for inequalities is more limited than it is for equations. Be sure to use it to multiply or divide only by numbers that you know are positive.

Q6 On page 4 the pans are balanced again. This time the objects are arranged in two identical stacks on each pan. What happens if you remove one complete stack from each pan? What arithmetic operation does this correspond to? This is Rule 3. Write it down.

Q7 Page 5 has unbalanced pans. Write down the inequality. Use the rules to get a single x on one pan and only numbers on the other. Write down the solution.

EXPLORE MORE

You can adjust the value of x if you want. Press the *Show x* button and then move the slider.

Q8 Page 6 has empty pans. Create your own problem by dragging objects onto the pans. Make sure this is a problem that can be solved without fractions. Save your problem, and ask a classmate to try it.

Graphing Inequalities in Two Variables

KEEP ME SATISFIED

Q1 The calculation $2x + 3$ stays greater than 5 as long as P is to the right of a vertical line through 1 on the x-axis.

Q2 The calculation $2y + 3$ stays greater than 5 as long as P is above a horizontal line through 1 on the y-axis.

Q3 The tracer is cyan (light blue) when the calculation is greater than 5, black when it's equal to 5, and magenta (light purple) when it's less than 5.

Q4 The graph of $2y + 3 = 5$ is the horizontal line with a y-intercept of 1. The points satisfying $2y + 3 > 5$ are all points above that line.

EXPLORE

Q5 When you move P straight down, the y-value decreases. In the expression, y is multiplied by -2, so the value of the $-2y$ term increases when y decreases. Similarly, when you move P to the right, the x-value increases, so the value of the $3x$ term also increases.

Q6 The statement $3x - 2y > 5$ is true when P is to the right of or below the line that is the graph of $3x - 2y = 5$. Because the value of the expression is 5 when P is on the line, and the value increases when you move P to the right or down, every point on this side of the line must correspond to an expression value greater than 5.

Q7 In the expression $x + 2y$, the value of the expression increases when you move P to the right or up, because the first term (x) increases when you increase x by moving P right, and the second term $(2y)$ increases when you increase y by moving P up. For this reason, the statement $x + 2y > -1$ is true when P is to the right of or above the line that is the graph of $x + 2y = -1$.

Q8 To satisfy the inequality $3x - y > 2$, the value of the expression must be greater than the value of the expression on the line $3x - y = 2$. The x term has a positive coefficient, so P must be to the right of the line; the y term has a negative coefficient, so P must be below the line. All points to the right of and below the line satisfy the inequality.

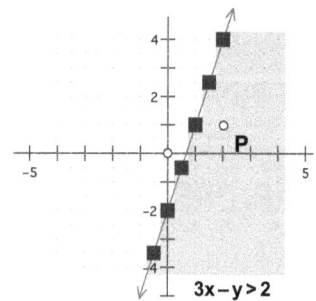

$3x - y > 2$

Q9 a.

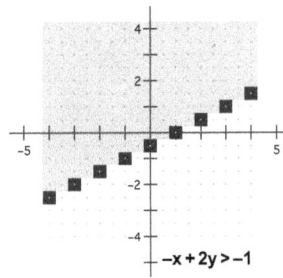

−x + 2y > −1

b.

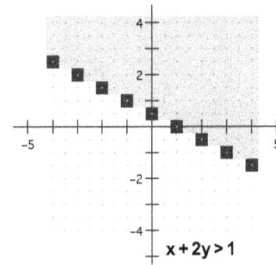

x + 2y > 1

c.

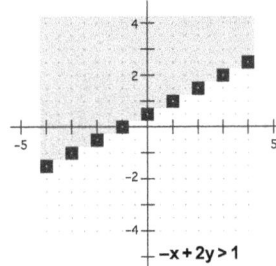

−x + 2y > 1

d.

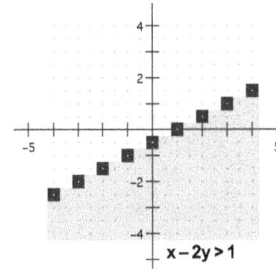

x − 2y > 1

Q10 The points that satisfy the inequality $y < x^2$ are the points below the parabola $y = x^2$.

Q11 The inequality $y > 3x - 5$ is true when y is above the line because y has a positive coefficient and is on the greater-than side of the inequality. It's true when x is left of the line because x has a positive coefficient and is on the less-than side of the inequality.

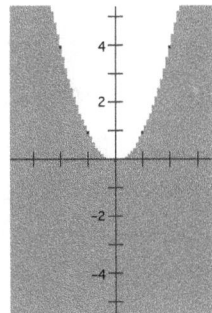

EXPLORE MORE

Q12 a.

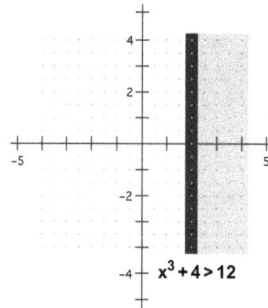

$x^3 + 4 > 12$

b.

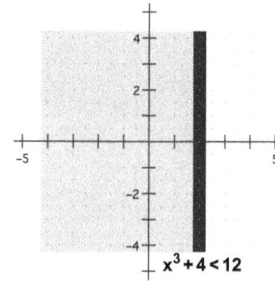

$x^3 + 4 < 12$

c.

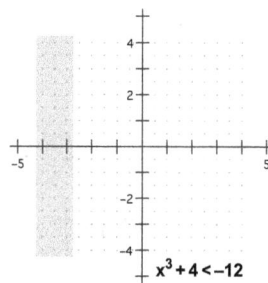

$x^3 + 4 < -12$

d.

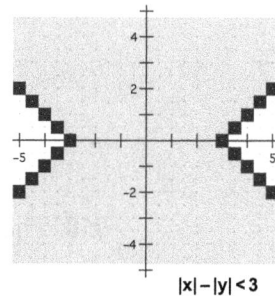

$|x| - |y| < 3$

e.

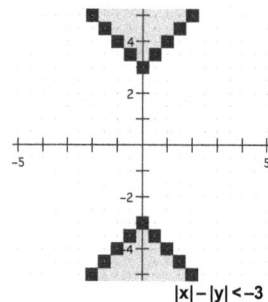

$|x| - |y| < -3$

f.

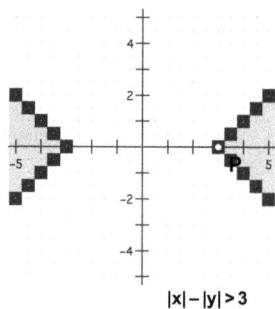

$|x| - |y| > 3$

Q13 After students sketch the graphs, comparing graphs a and b and graphs d and f helps them to see that the direction of the inequality sign determines whether the cyan region or the magenta region represents the solution.

The curve(s) representing equality divide the plane into two regions. If you reverse the sign of the inequality, you change which region is the solution.

Q14 Each of these inequalities has a factored polynomial on the left side and zero on the right. To determine whether the inequality is satisfied for a given point, you only need to know the sign of the left side. Find that by determining the sign of each factor. The boundaries are vertical lines at the roots. Every time you cross one, one of the factors changes sign, and that changes the sign of the whole polynomial. Watch out for double roots; none are used in these examples.

Q15 This inequality has an interesting shape. Students may want to adjust the grid extent to view more of it.

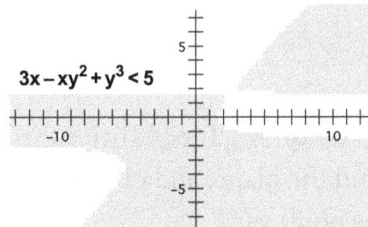

The related inequality $3x - xy^2 + y^3 > 5$ has the inequality sign reversed, so the solution is the complementary region (the region not colored in the diagram). The inequality $-(3x - xy^2 + y^3) < -5$ has the same shape as $3x - xy^2 + y^3 > 5$, because the direction of the inequality is reversed, and the signs of both sides have also been reversed.

If you change the inequality to an equation and solve for x in terms of y, you can graph the result. (In the Function Calculator, set the Equation form to be $x = f(y)$.) You'll need to increase the number of samples in the function plot to see the complete graph.

WHOLE-CLASS PRESENTATION

Use the sketch **Graphing Inequalities Present.gsp** to present this activity to the class. The presentation follows the steps and questions of the student activity, but is streamlined to make it easy to present. Each page of the presentation sketch has directions and buttons to help you present the activity. Use pages 1 through 7 to present the main part of the activity. Pages 8 through 10 correspond to the Explore More questions.

Graphing Inequalities in Two Variables

Imagine yourself as a point in the coordinate plane, free to wander. In this activity you'll travel to different locations, learning how to keep inequalities satisfied.

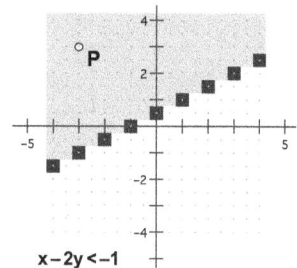

$x - 2y < -1$

KEEP ME SATISFIED

1. Open **Graphing Inequalities.gsp.** Move point P around the plane, and observe how the coordinates of the point vary.

Q1 Where can you move the point so that the calculation $2x + 3$ stays greater than 5? Describe these locations.

> To edit the calculation, double-click it. Click the y measurement in the sketch to enter it into the calculation.

Q2 Edit the calculation to $2y + 3$. Now where can you move the point so that the calculation of $2y + 3$ stays greater than 5? Describe these locations.

Q3 Press *Show Tracer* and move point P around again, this time anywhere you like. What happens to the tracer when the calculation $2y + 3$ is greater than 5, when it's equal to 5, and when it's less than 5?

> Although the inequality $2y + 3 > 5$ has only one variable, we're still interested in all the coordinate points, which have two variables.

A graph of an inequality in two variables is the collection of all the points in the plane that *satisfy* the inequality. In this context, satisfy means "make true." For example, any time point P is in a position where $2y + 3$ is greater than 5, the inequality $2y + 3 > 5$ is true.

Q4 What points satisfy the equation $2y + 3 = 5$? How might knowing this help you describe the points that satisfy the inequality $2y + 3 > 5$?

EXPLORE

> To change the calculation, double-click it and then use the Calculator. Enter values of x or y by clicking the x or y measurements in the sketch.

2. Erase the traces, hide the tracer, and change the calculation to $3x - 2y$.

Q5 Given any position of point P, if you move it straight down (without moving it left or right), the value of $3x - 2y$ increases. Explain why. What happens to the calculation if you move P to the right without moving it up or down? Why?

3. Show the tracer and move point P around the plane.

Q6 Describe the locations of point P where $3x - 2y > 5$ is true. Use your answer to Q4 to explain why any point to the right of (or below) the line $3x - 2y = 5$ is a point that satisfies the inequality $3x - 2y > 5$.

4. On page 2, drag point P and notice how it moves between closely spaced points. Show the tracer and drag P again. The tracer changes color, showing for each location of point P whether $x + 2y > -1$ is true.

Q7 Use your answers to Q5 and Q6 to explain why one side of the line $x + 2y = -1$ contains *all* the points where $x + 2y > -1$.

Q8 Without using Sketchpad, use your answers to Q4 through Q6 to predict what points in the plane will satisfy the inequality $3x - y > 2$. Write out your thinking in a sentence, and draw a sketch of the graph on your paper. Then check your answer by editing the calculation and moving point P.

Q9 For each inequality below, sketch on your paper your prediction of the graph. Then use Sketchpad to check your result.

a. $-x + 2y > -1$ b. $x + 2y > 1$

c. $-x + 2y > 1$ d. $x - 2y > 1$

Q10 Use page 3 to find points that satisfy the inequality $y < x^2$. How can you determine, without Sketchpad, which points in the plane satisfy $y < x^2$?

Q11 Answer Q10 for the inequality $y > 3x - 5$, again without using Sketchpad.

EXPLORE MORE

To see the entire graph quickly, select both *P* and the tracer square, and choose **Construct | Locus.** Then use **Edit | Properties | Plot** to increase the number of samples so that the locus fills the entire region.

Q12 Use page 4 to find the points in the plane that satisfy the inequalities below. Draw a sketch of each solution. Think about how you could go about predicting the points that satisfy the inequality without Sketchpad available.

a. $x^3 + 4 > 12$ b. $x^3 + 4 < 12$

c. $x^3 + 4 < -12$ d. $|x| - |y| < 3$

e. $|x| - |y| < -3$ f. $|x| - |y| > 3$

Q13 When you're done, compare the solutions to a, b, and c with each other, and the solutions to d, e, and f with each other. What conclusions, if any, can you draw about reversing the inequality sign?

Q14 Use page 5 to find the points that satisfy the inequalities below. How could you predict the points that satisfy the inequality without Sketchpad available?

a. $(x - 2)(x + 1) < 0$ b. $(x - 2)(x + 1)(x - 4) < 0$

Q15 Use page 6 to find points that satisfy the inequality $3x - xy^2 + y^3 < 5$. (Make sure you drag point P to a variety of locations.) What points would satisfy the inequality $3x - xy^2 + y^3 > 5$? Why? What points would satisfy the inequality $-(3x - xy^2 + y^3) < -5$? Why? Check your answers with the sketch and explain the results that you see.

Graphing Systems of Inequalities

A SINGLE INEQUALITY

The graphing tools used in this activity require inequalities (like $y > x - 2$ or $x \leq y^2 - 5$) that have been "solved" for either x or y. Most of the inequalities in this activity are already in this form. In Q7 students are asked to do the "solving" themselves. Use Q7 as a chance to review the properties of inequality needed to convert complicated expressions to the form required by the tools.

Q1 The graph crosses the y-axis at -2 and the x-axis at 1.33 and goes up and to the right with a slope of 1.5.

Q2 The inequality says that y must be less than the value of the function just graphed, so the graph of the inequality must be below the line. It extends infinitely down and to the right.

4. If the graph appears bounded by a solid line rather than a dashed line, the student probably failed to delete the original line in step 2. Undo the custom tool, delete the line, and then use the custom tool again.

5. To avoid the need to increase the number of samples for every graph, hold down the Shift key and choose **Edit | Advanced Preferences.** On the Locus panel, change the number of point-locus samples from 300 to about 1000. (The exact value required depends on the size of the sketch window.)

Q3 The line itself is not included in the graph because y cannot be equal to the value of the function. For this reason it should appear as a dashed line.

SYSTEMS OF INEQUALITIES

Q4 Answers will vary; the important thing is that students make a conjecture before constructing the graph. (It's actually the region of the plane above and to the right of the line $x = -y + 1$.)

Q5 The graph is a portion of the plane bounded by the two lines, shown by the darkest area in the figure. This graph is infinite in extent.

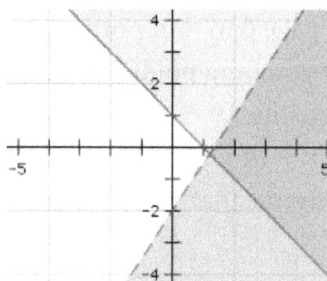

Q6 The area where the two shades overlap is the solution to the system. The graphs in problem e do not overlap, so it has no solution. In problem h, have students set the angle units to radians before graphing.

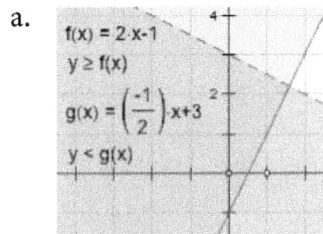

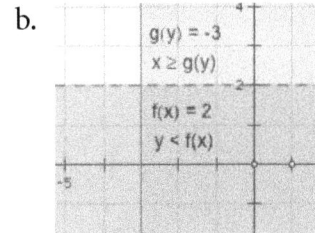

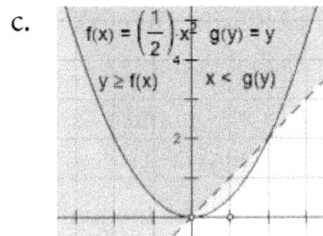

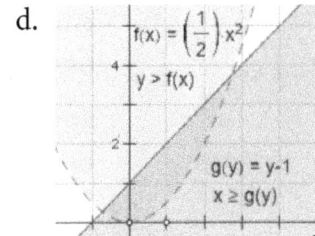

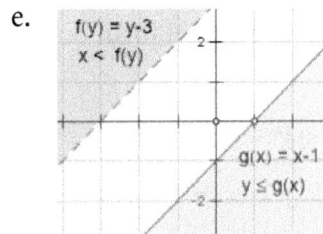

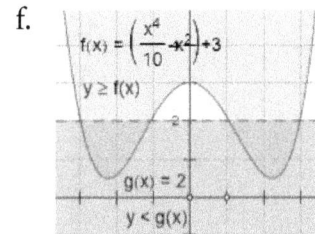

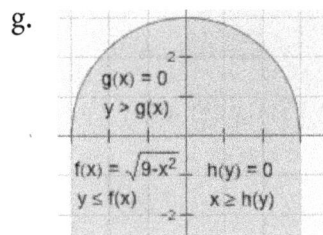

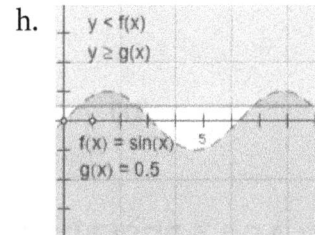

a.

$f(x) = 2 \cdot x - 1$
$y \geq f(x)$

$g(x) = \left(\dfrac{-1}{2}\right) \cdot x + 3$
$y < g(x)$

b.

$g(y) = -3$
$x \geq g(y)$

$f(x) = 2$
$y < f(x)$

c.

$f(x) = \left(\dfrac{1}{2}\right) x^2$ $g(y) = y$
$y \geq f(x)$ $x < g(y)$

d.

$f(x) = \left(\dfrac{1}{2}\right) x^2$
$y > f(x)$

$g(y) = y - 1$
$x \geq g(y)$

e.

$f(y) = y - 3$
$x < f(y)$

$g(x) = x - 1$
$y \leq g(x)$

f.

$f(x) = \left(\dfrac{x^4}{10} - x^2\right) + 3$
$y \geq f(x)$

$g(x) = 2$
$y < g(x)$

g.

$g(x) = 0$
$y > g(x)$

$f(x) = \sqrt{9 - x^2}$ $h(y) = 0$
$y \leq f(x)$ $x \geq h(y)$

h.

$y < f(x)$
$y \geq g(x)$

$f(x) = \sin(x)$
$g(x) = 0.5$

EXPLORE MORE

Q7 a.   The first inequality can be expressed as $x > 2y - 2$ or as $y < 1/2x + 1$. The second inequality can be expressed as $y \geq x/2 + |x|$.

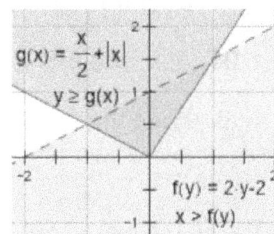

$g(x) = \dfrac{x}{2} + |x|$
$y \geq g(x)$

$f(y) = 2 \cdot y - 2$
$x > f(y)$

b. The first inequality can be expressed as $x \geq y^2 - 2y + 2$. The second inequality can be expressed as $x < 3$.

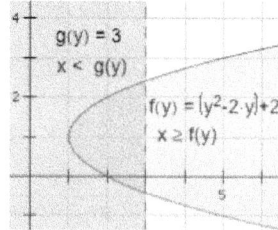

Q8 For some of the inequalities, it's possible to express the answers with either x or y on the left side. Here is one answer for each graph, with y on the left side of each inequality.

a. $y < 0.5x + 1$, $y > x - 1$, and $y \geq -2x - 1$
b. $y > |x|$ and $y \leq -0.5x + 1$

Q9 It's not possible to rewrite the inequality $y > |x|$ in a simple form with x on the left side. Here are the results for rewriting each of the other inequalities, with the original form on the left and the rewritten form on the right:

$$y < 0.5x + 1 \qquad\qquad x > 2y - 2$$

$$y > x - 1 \qquad\qquad x < y + 1$$

$$y \geq -2x - 1 \qquad\qquad x \geq -0.5y - 0.5$$

$$y \leq -0.5x + 1 \qquad\qquad x \leq -2y + 2$$

Q10 Answers will vary. Consider asking a few students to show and explain their answers to the class.

WHOLE-CLASS PRESENTATION

Use **Systems of Inequalities Present.gsp** to present this activity to the class. The presentation sketch first reviews graphing the single inequality $y \leq 3/2x - 2$.

1. On page 1, remind students that the first step in graphing an inequality is to graph the related equation. Use the button to show the function plot.

Q1 Ask students to predict where the graph of the inequality will be.

2. Hide the function plot and show the inequality.

Q2 Ask students why the boundary of the graph is a dashed line. Review the convention that a boundary that is part of the graph is shown as a solid line, and a boundary that is not part of the graph is shown dashed.

3. Go to page 2 to add the second inequality.

Q3 Ask students to predict where the graph will be. Have them explain how the graph of the inequality is related to the graph of $y = -x + 1$. Use Sketchpad's **Line** tool to draw an arbitrary line and have students tell you where to drag it to indicate the boundary.

Q4 Once students agree on where to place the boundary, ask them which side of the boundary the graph will be on. Use the tracer provided to check the result.

4. Press *Show y >= −x + 1* to show the result.

Q5 Ask students, "Why is this boundary solid?"

Q6 Ask students, "Where are the points that satisfy both inequalities? What is the solution region?"

5. Go to page 3, which contains a different system of inequalities.

Pages 3 through 8 contain various systems of inequalities for students to practice on. Use as many of these pages as are appropriate. Be sure to have students make a prediction before showing them the result on the computer.

Q7 Ask, "What will the first inequality look like?"

6. Use the buttons to show first the related function and then the inequality.

7. Similarly, ask about and show the second function and inequality, and then discuss the solution.

Pages 9 and 10 contain graphs of systems of inequalities. (These are the graphs from Q8 in the activity.) Use either page as a challenge to students to come up with a system of inequalities that matches the graph. Use the custom tools in **Inequality Tools.gsp** to verify student answers.

Graphing Systems of Inequalities

When you solve a system of equations graphically, you usually get a single point as the solution. In this activity you'll solve a system of inequalities graphically, and see what the solution of such a system looks like.

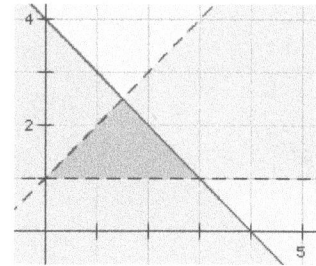

A SINGLE INEQUALITY

Begin by graphing the equation $y = \frac{3}{2}x - 2$ and the related inequality $y < \frac{3}{2}x - 2$.

1. In a new sketch, graph $y = \frac{3}{2}x - 2$ by choosing **Graph | Plot New Function** and entering the equation into the New Function dialog box.

Q1 What is the slope of the line and what are its *x*- and *y*-intercepts?

Q2 Based on the graph of the *equation*, what do you think the graph of the *inequality* will look like?

> To delete the graph, select it and press the Delete or Backspace key on your keyboard.

2. Delete the function plot you just created, but leave the function itself; you will use it again in a moment.

To graph the inequality, you will use a custom tool from **Inequality Tools.gsp.**

3. Open **Inequality Tools.gsp.** Then switch back to your original sketch. (As long as **Inequality Tools.gsp** is open, you can use the tools it contains.)

> Switch back to the **Arrow** tool after the graph appears.

4. Press and hold the **Custom** tool icon, and choose **Inequality Tools | y < f(x)** from the menu that appears. Click the function $f(x)$ in your sketch. A graph of the inequality appears.

> You can also right-click (Windows) or control-click (Mac) the strips to get to the Properties dialog box.

5. If the graph of the inequality appears as strips, you can change the Plot Properties of this object so that it appears as a solid area. Select the strips, choose **Edit | Properties,** and use the Plot panel of the Properties dialog box to increase the number of samples. (Approximately 400 should be enough.)

Q3 Is the boundary of the inequality graph solid or dashed? Explain why its appearance makes sense in terms of the inequality.

SYSTEMS OF INEQUALITIES

Next you'll graph a second inequality ($x \geq -y + 1$) on the same coordinate system.

> This function should be labeled *g*. If not, choose **Display | Label Function** to change its label to *g*.

6. Create another function by choosing **Graph | New Function.** This inequality expresses *x* in terms of *y*, so use the Calculator's Equation pop-up menu to choose **x = f(y).** Then enter the equation $g(x) = -y + 1$.

Q4 What do you think the graph of $x \geq -y + 1$ will look like? Write down your guess before you construct the graph.

Remember to switch back to the **Arrow** tool once the graph appears.

7. To create the graph, choose the custom tool **Inequality Tools | x >= f(y)**. Click the tool on the function $g(y)$ in your sketch.

Q5 Which shaded area contains points that satisfy both inequalities? Construct a point in that area, measure its coordinates, and use algebra to confirm that the point satisfies both inequalities.

In addition to the colors in the color menu, you can choose **Other...** to specify a different shade or tint.

Q6 For each system of inequalities below, add a new page to your document (using **File | Document Options**) and construct a graph of the system. If the graphs of two inequalities appear in the same color, change the color of one of the graphs so you can easily see the area of overlap.

a. $y \geq 2x - 1, y < -\frac{1}{2}x + 3$ b. $y < 2, x \geq -3$

For problem h, use **Edit | Preferences | Units** and change the angle units to radians.

c. $y \geq \frac{1}{2}x^2, x < y$ d. $y < \frac{1}{2}x^2, x \geq y - 1$

e. $x < y - 3, y \leq x - 1$ f. $y > \frac{x}{10} - x^2 + 3, y < 2$

g. $x \leq \sqrt{9 - y^2}, y > 1, x \geq -1$ h. $y < \sin x, y \geq 0.5$

EXPLORE MORE

Q7 Change these inequalities so that each variable appears on only one side of the inequality, making sure to keep the properties of inequality in mind. Then graph the system and describe the solution.

a. $2y - x < 2, |2x| \geq 2y - x$ b. $y^2 \leq x + 2y - 2, x - 3 < 0$

Some inequalities can be expressed with either x or y on the left side, so more than one answer may be possible for a particular graph.

Q8 Study each graph below and decide on a system of inequalities that will produce it. Check each answer on a new page of your document.

a.

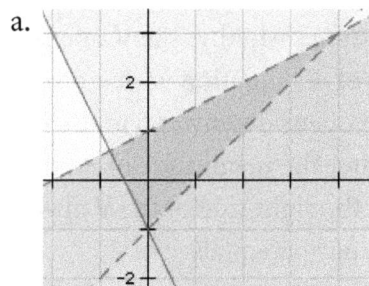

b.

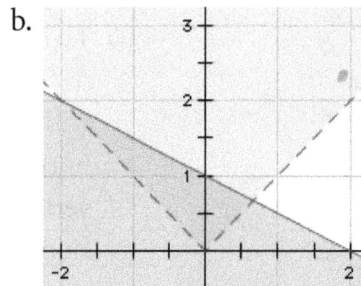

Q9 You probably expressed your answers to the preceding question in a form with y alone on the left side of each inequality. Where possible, rewrite your answers so that x is alone on the left side.

Q10 Use inequalities to graph an interesting shape in Sketchpad. Then challenge a friend to guess the inequalities you used.

Absolute Value Functions

In addition to familiarizing students with various transformations of the absolute value graph, there are two other reasons for doing this activity:

First, it reinforces students' understanding of the point-slope form of lines (with the role of point (h, k) becoming even more apparent).

Second, it prepares students for the vertex form of parabolas.

It's important for students to actually predict what the absolute value graphs will look like before they graph them. Just by plotting a few points—such as $(-2, 2)$, $(0, 0)$, and $(2, 2)$—they can gain insight into why these graphs appear as they do.

SKETCH AND INVESTIGATE

Q1 The graphs are identical to the right of $x = 2$. To the left of $x = 2$, the graphs are reflections on the x-axis. Another way of describing this is that the part of $y = 2x - 4$ that was below the x-axis has been flipped above the x-axis. The graph of $y = 2x - 4$ is a line whereas the graph of $y = |2x - 4|$ is shaped like a **V**. The range of $y = 2x - 4$ is all real numbers whereas the range of $y = |2x - 4|$ is all real numbers greater than or equal to zero.

Q2 This graph has the same parent function ($y = 2x - 4$), but a different vertex: the point $(0, -4)$. The purpose of this question is to spur students to think about how the numbers in the equation relate to the position of the vertex.

A FAMILY OF ABSOLUTE VALUE GRAPHS

Q3 The sign of m determines whether the **V** opens upward or downward or is flat. The graph of an equation with a positive m opens upward; with a negative m, it opens downward; and with $m = 0$, it is straight and horizontal. The greater the magnitude of m, the smaller the angle at the vertex. The slope of the right side of the **V** always equals m while the slope of the left side always equals $-m$.

Q4 The coordinates of the vertex are (h, k).

Q5 a. $y = 2|x + 1| + 2$ b. $y = -1.5|x - 2| + 3$

c. $y = 3|x + 3| - 1$ d. $y = (2/3)|x| - 8/3$

e. $y = -1|x - 2| + 4$ f. $y = -2.5|x - 2| + 5$

Q6 The graph does not have *x*-intercepts if parameters *m* and *k* have the same sign. For example, if *k* is positive, then the vertex is above the *x*-axis. If *m* is also positive, then the graph opens upward, so it can never reach down to the *x*-axis.

If *m* and *k* have opposite signs, then there are two *x*-intercepts. If $k = 0$, then the vertex itself is an *x*-intercept.

EXPLORE MORE

Q7 Consider the graphs of two equations: $y = f(x)$ and $y = |f(x)|$. Where the first graph is below the *x*-axis, it will be reflected across (above) the *x*-axis in the second graph. Where the first graph is above the *x*-axis, the second graph will be identical.

1. Open **VGraph Present.gsp.** Press *Show y = x.* After a short pause, press *Show y = | x |.*

Q1 It looks like the graphs coincide on the right side. Geometrically, what is their relationship on the left? (They are reflections on the *x*-axis.)

There is also a reflection on the *y*-axis. If anyone mentions that, acknowledge it and verify that reflection too.

2. To verify this, construct a point on the left side of either graph. Double-click the *x*-axis to mark it as a mirror. Select the point and choose **Transform | Reflect.** Drag the first point to show that the image follows the other graph.

3. On page 2, press *Show y = mx + b.* Drag the two sliders to show how they control the graph.

Q2 Ask, "What will the graph of $y = | mx + b |$ look like?" Discuss predictions and show the graph.

Q3 Give *m* a positive value for this question. What is the slope on the right part of the absolute value graph? (*m*) What is the slope on the left side? ($-m$)

4. The slope on the right side should be clear because of the coinciding graphs. To check the slope on the left side, construct a line segment with both endpoints on the left side of the absolute value graph. Select the segment and choose **Measure | Slope.**

5. On page 3 is the graph of $y = m | x |$, with $m = 1$ to begin.

There is a button that will show *m* as a fraction, reinforcing the slope concept.

Q4 Before moving the slider, ask for predictions. Then vary *m* and ask for observations as you make *m* bigger, smaller, negative, and zero. (As with multiplying any other function by a constant, *m* defines the ratio for a vertical stretch. In this case an *m* with larger magnitude makes the vertex angle smaller. If *m* is positive, it opens upward; if negative, downward. When $m = 0$, the graph coincides with the *x*-axis.)

6. Page 4 contains three sliders and the graph of $y = m | x - h | + k$.

Q5 What are the effects of changing the parameters? Parameters *h* and *k* define horizontal and vertical translation. Parameter *m* has the same effect as before.

Q6 Identify the vertex. What are the coordinates of this point? (*h*, *k*) To emphasize this, press *Reset*, and then drag the *h* and *k* sliders one at a time. Give students the coordinates of the vertex and one other point, and challenge them to derive corresponding settings for *m*, *h*, and *k*.

7. Page 5 has the graph of a general function along with its absolute value. Edit the definition of *f(x)* and have students predict the shape of its absolute value. Challenge them with functions they have never seen. Given the root function graph, they should still be able to predict the shape of the absolute value.

Absolute Value Functions

The absolute value of a number is how big the number is, regardless of whether it's positive or negative. Some examples should make this clearer.

The absolute value of -5 is 5,   *or* $|-5| = 5$.

The absolute value of 5 is 5,   *or* $|5| = 5$.

The absolute value of 0 is 0,   *or* $|0| = 0$.

As you can see, the absolute value of a number is always a positive number or zero.

But what happens when you graph an equation involving the absolute value function, such as $y = |2x - 4|$? In this activity you'll find out.

SKETCH AND INVESTIGATE

Choose **Graph | Plot New Function** to open the New Function calculator. Then type x and click OK.

1. In a new sketch, plot the equation $y = x$.

2. Think about what the graph of $y = |x|$ might look like. Plotting points by hand or discussing the question with classmates might help. Make a rough sketch of your guess on scratch paper.

Choose **Plot New Function** again. Then choose **abs** from the Functions pop-up menu, enter x, and click OK.

3. Plot the equation $y = |x|$. How does it compare with your prediction?

4. Repeat steps 1–3 with the equations $y = 2x - 4$ and $y = |2x - 4|$. Make sure to draw a prediction of what you think the second equation will look like before plotting it in Sketchpad.

Q1 Describe how the graphs of $y = 2x - 4$ and $y = |2x - 4|$ compare. Discuss their shapes, their ranges, and other features you notice.

Q2 Predict what the graph of $y = |2x| - 4$ will look like. After making your prediction, plot it. Then describe the differences between this graph and the other two.

A FAMILY OF ABSOLUTE VALUE GRAPHS

As you explore this family, keep in mind a related family: lines in point-slope form $y = m(x - h) + k$.

Now that you have an idea of what absolute values can do to the graphs of particular functions, you'll explore a *family* of graphs: $y = m|x - h| + k$.

5. Open **VGraph.gsp.** You'll see the graph of an equation in the form $y = m|x - h| + k$, and sliders for the parameters m, h, and k.

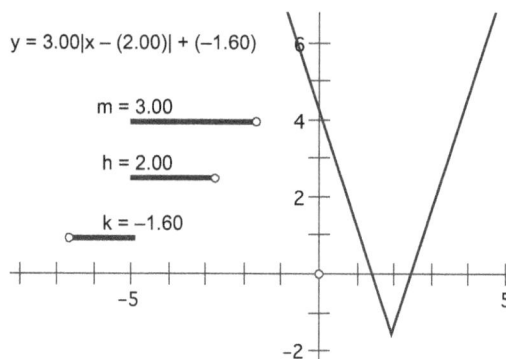

$y = 3.00|x - (2.00)| + (-1.60)$

$m = 3.00$

$h = 2.00$

$k = -1.60$

Q3 Adjust slider m and observe the effect this has on the **V**-graph. Try different values for the slope—large, small, positive, negative, and zero. Summarize the role m plays in the equation $y = m|x - h| + k$.

Q4 Changing m moves the entire **V**-graph except for one point. This point is called the *vertex*. Adjust the sliders for h and k. How does the location of the vertex relate to the values of h and k?

Q5 Write an equation in the form $y = m|x - h| + k$ for each of the **V**-graphs described below. Check each answer by adjusting the m, h, and k sliders.

 a. Vertex at $(-1, 2)$; contains the point $(0, 4)$

 b. Vertex at $(2, 3)$; contains the point $(4, 0)$

 c. Vertex at $(-3, -1)$; same shape as $y = 3|x - 2| + 5$

 d. x-intercepts at $(4, 0)$ and $(-4, 0)$; contains the point $(1, -2)$

 e. x-intercepts at $(6, 0)$ and $(-2, 0)$; contains the point $(5, 1)$

 f. Same vertex as $y = 3|x - 2| + 5$; contains the point $(0, 0)$

Q6 Not all graphs in this family have x-intercepts. How can you tell whether a function has x-intercepts just by looking at the function parameters?

EXPLORE MORE

To enter sin(x), choose **sin** from the Calculator's Functions pop-up menu. For this function's graph to appear properly, Sketchpad's angle units must be set to radians. Use **Edit | Preferences | Units** to check this setting.

Q7 Plot the following two pairs of equations:

 a. $y = x^2 - 1$ and $y = |x^2 - 1|$

 b. $y = \sin x$ and $y = |\sin x|$ (Don't worry if you're unfamiliar with the sin function.)

Write a short paragraph summarizing what happens when you plot a function and its absolute value.

Exponential Functions

PROPERTIES OF THE GRAPH

Q1 Both a and b are nonzero. Therefore, ab^x is also nonzero for any real x, so y cannot be zero, and there is no x-intercept.

For the y-intercept, substitute zero for x in the equation $y = ab^x$.

$$y = ab^0 = a$$

The y-intercept is a.

Q2 For $b > 1$, $f(x)$ tends to zero on the left. For $0 < b < 1$, $f(x)$ tends to zero on the right. The value of a has no influence on this property.

Q3 If a were equal to zero, the function would be the constant function $f(x) = 0$.

If b were equal to zero, the function would be zero for all positive x, and it would be undefined for all other values of x.

If b were less than zero, b^x would not be continuously defined over all real exponents x.

If b were equal to one, this would be another constant function, $f(x) = a$.

Q4 Limited resolution may prevent students from making the difference exactly 1.00. It's sufficient for them to make it as close to that value as they can. The ratio $y_Q/y_P = 1.30$. This is the same as parameter b.

$$\frac{y_Q}{y_P} = \frac{f(x_Q)}{f(x_P)} = \frac{ab^{x_Q}}{ab^{x_P}} = b^{x_Q - x_P} = b^1$$

Here it does not matter where on the graph points P and Q are, so long as $x_Q - x_P = 1$.

Q5 If $x_Q - x_P = 2.5$, then $y_Q/y_P = 1.30^{2.5} \approx 1.9$. This follows from the same reasoning as in the previous answer.

$$\frac{y_Q}{y_P} = \frac{f(x_Q)}{f(x_P)} = \frac{ab^{x_Q}}{ab^{x_P}} = b^{x_Q - x_P} = b^{2.5}$$

DOUBLING PERIOD AND HALF-LIFE

8. There may be some confusion regarding the term *effective annual yield*. The actual rate is about 5.8%, but since it is compounded continuously, at the end of each year, the investment will be worth 6% more than it

was at the beginning of the year. Even if students do not yet grasp the concept, they can continue with the given function definition.

Q6 The doubling period is about 11.90 years.

Q7 To find the half-life, students should arrange the points so that $y_Q/y_P = 0.50$. When that happens, $x_Q - x_P \approx 30$, so cesium has a half-life of about 30 years. In this case changes in the scales of the axes can cause quite a lot of variation in the answers.

Q8 To find a doubling period, with the ratio $y_Q/y_P = 2.00$, point Q would have to be to the left of P. In that case, $x_Q - x_P \approx -30$.

This answer fits with the half-life answer. It stands to reason that if it takes 30 years for half of the cesium to decay, then 30 years ago, there was two times as much. This explains the negative doubling period.

1. Begin by showing the general form of this exponential function:

$$f(x) = ab^x, \quad \text{where } a \neq 0,\ b > 0,\ \text{and } b \neq 1$$

Press *Show Slider Controls* to reveal buttons that you can use to set the parameters to various precise values.

2. Open **Exponential Present.gsp.** This is a graph of the function with sliders controlling the values of *a* and *b*. Drag each slider in turn so that students can see the effects on the graph.

Q1 Ask students for the *x*- and *y*-intercepts. [There is no *x*-intercept, and the *y*-intercept is equal to *a*.] Challenge them to verify these facts by alternately setting *y* and *x* to zero in the equation $y = ab^x$.

Q2 Sometimes the function approaches zero on the left side, and sometimes on the right. What determines which side it is? [It's on the left when $b > 1$ and on the right when $0 < b < 1$.]

Q3 In the general form of the function, there are three restrictions on the parameters *a* and *b*. Why? Show students what happens when $a = 0$, $b < 0$, $b = 0$, or $b = 1$.

3. On page 2 there are two points on the graph, *P* and *Q*. You can drag *P* freely, but point *Q* is a fixed distance to the right of point *P*. That distance is determined by the slider labeled Δ. At the bottom of the screen are measurements showing the difference of the *x*-coordinates and the ratio of the *y*-coordinates.

4. Drag point *P* to show that the ratio of the *y*-coordinates remains constant when the difference in the *x*-coordinates (Δ) is constant.

On pages 3 and 4, the *a* and *b* sliders have been replaced with parameters in order to make it easier to enter precise values.

Q4 What will the ratio be when $\Delta = 1.00$? [It should equal *b*.] Challenge students to predict this before showing it. Then have them prove that $\dfrac{f(x_P + 1)}{f(x_P)} = b$.

5. Page 3 has a graph showing the growth of an investment of $100 with an effective annual yield of 6%.

Pages 3 and 4 have rectangular grids. If students have not used that feature yet, this would be a good opportunity to show them the advantages of using different scales on the axes.

Q5 With this investment, how long would it take to double your money? If the money is doubled between *P* and *Q*, then the ratio of their *y*-coordinates will be 2. Drag the Δ slider until the ratio is 2.00. [The doubling period is the difference in the *x*-coordinates, about 11.9 years.]

6. The graph on page 4 shows the radioactive decay of 80 g of cesium. The *x*-scale is in years.

Q6 What is the half-life of cesium? This question is similar to the previous one. Give students time to figure out that they need to adjust the difference in the *x*-coordinates so that the ratio is 0.50. [The result is a half-life of about 30 years.]

Exponential Functions

There is a connection between population growth, radioactive decay, musical scales, and compound interest. They seem to have little in common, but you can model any of them using an exponential function.

An exponential function has the general form $f(x) = ab^x$, where $a \neq 0$, $b > 0$, and $b \neq 1$.

PROPERTIES OF THE GRAPH

Before using an exponential function to model a real-world problem, take some time to familiarize yourself with the graph.

To create a parameter, choose **Number | New Parameter.**

1. In a new sketch, create parameters a and b.

2. Graph the function $f(x) = a \cdot b^x$ by choosing **Graph | Plot New Function.** Click the parameters in the sketch to enter them into the function.

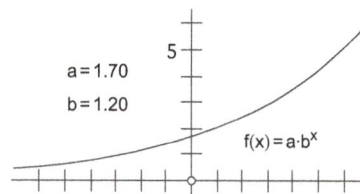

$a = 1.70$
$b = 1.20$
$f(x) = a \cdot b^x$

To change the value of a parameter, either double-click it or select it and press the + or − key. (To change the size of the steps for the + or − keys, select the parameter and choose **Edit | Properties | Parameter.** Change the Keyboard Adjustments value to 0.1 unit.)

3. The graph is plotted on the screen. Change the values of the parameters, and observe the resulting changes in the graph. Try several values for each parameter.

Q1 What are the x- and y-intercepts of the graph? Explain how the intercepts are related to parameters a and b.

Q2 The value of $f(x)$ tends to get close to zero either on the left side or on the right. What parameter values determine which side it is?

Q3 In the general form of the exponential function, there are three constraints ($a \neq 0$, $b > 0$, and $b \neq 1$). Explain the reason for each of these constraints.

Next you'll investigate how the function behaves by comparing the coordinates of two points on the graph.

4. Change the parameters so that $a = 2.00$ and $b = 1.30$.

5. Construct two points on the function graph. Label them P and Q.

6. Select both points and choose **Measure | Abscissae (x).** Select the points again and choose **Measure | Ordinates (y).**

7. Calculate the values $x_Q - x_P$ and y_Q/y_P.

You can change the scale of the axes to give you more precise control over the positions of the points.

Q4 Drag point Q one unit to the right of P, so that the difference $x_Q - x_P$ is as close to 1.00 as you can make it. What is the value of the ratio y_Q/y_P? Drag point P to a different

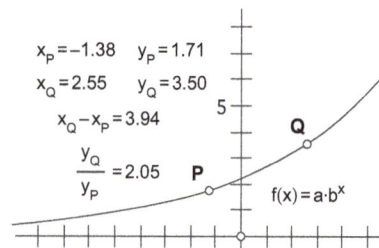

$x_P = -1.38$ $y_P = 1.71$
$x_Q = 2.55$ $y_Q = 3.50$
$x_Q - x_P = 3.94$
$\dfrac{y_Q}{y_P} = 2.05$
Q
P
$f(x) = a \cdot b^x$

position on the graph, and again drag Q so it's one unit to the right. What is the value of the ratio y_Q/y_P? Why do you get this result?

Q5 Now use a difference other than 1. Drag the points so that $x_Q - x_P$ is approximately 2.5. What is the ratio? Drag them to another position, but with the x difference still equal to 2.5. What is the ratio now? Explain.

DOUBLING PERIOD AND HALF-LIFE

Exponential functions can be used to solve a number of real-life problems. First change the function so that it shows the value of $100 invested at an effective annual yield of 6%.

> Effective annual yield is not the same thing as interest rate. That's another topic.

8. Edit parameters a and b so that the function has the definition $f(x) = 100(1.06)^x$. This shows the value of $100 invested at an effective annual yield of 6%. (The x variable is in years, and y is in dollars.)

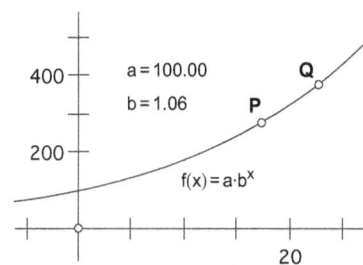

> Setting the grid to rectangular allows you to adjust each axis independently of the other.

9. At first you can't see the graph because the y-axis doesn't go up to 100. To adjust the axes, choose **Graph | Grid Form | Rectangular Grid.** Then drag tick mark numbers on each axis so that you can see the results for the first 25 years.

> To see more decimal places in a parameter, select the parameter and choose **Edit | Properties | Value.** Change the Precision setting.

Q6 How long will it take to double your money? Drag the points so that the ratio is 2.00. What is the difference in their x-coordinates? This number is called the *doubling period*.

An exponential function can also be used to model the decay of radioactive cesium.

10. To model the decay of 80 g of cesium, change the function definition to $f(x) = 80(0.977)^x$. Adjust the axes appropriately. The value of x is still in years, but the value of y is in grams.

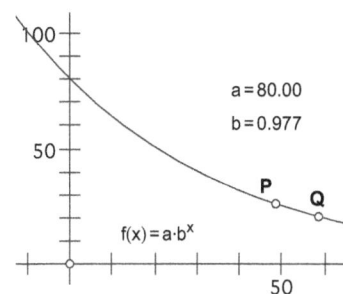

Q7 If you start with 80 g, you will have less cesium every year. How long would it take to lose half of it? Explain how you found the answer. This number is called the *half-life* of cesium.

Q8 Although cesium decays, as opposed to growing, you can still calculate its doubling period. Drag the two points until you find a position where the ratio is 2.00. What is the difference in the x-coordinates? Explain how this verifies your answer to Q7.

A few students may get through this quickly, but most will need a lot of time. It will be especially difficult for students who have poor computer skills. With so many expressions to enter, they may tend to make mistakes with order of operations. It is important not to rush them. Encourage them to work together.

VERTEX TO STANDARD

4. It will be helpful to students if you can demonstrate ahead of time how to use hot text to insert values into a caption. In your demonstration include the technique used in step 5, calculating the value of $-h$ before inserting it as an addend.

Q1 $2x^2 - 32x + 133$

STANDARD TO VERTEX

Q2 $h = -\dfrac{b}{2a}$ $k = \dfrac{4ac - b^2}{4a}$

Q3 When $a = 0$, the formulas for both h and k are undefined. However, it is also true that when $a = 0$, the standard form function is linear, not quadratic. In fact, you can rewrite any quadratic function from standard form to vertex form.

OTHER CONVERSIONS

Factored to Standard

Q4 $b = -a(r_1 + r_2)$ $c = ar_1 r_2$

Standard to Factored

Q5 $r_1 = \dfrac{-b - \sqrt{b^2 - 4ac}}{2a}$ $r_2 = \dfrac{-b + \sqrt{b^2 - 4ac}}{2a}$

Q6 The condition $a \neq 0$ occurs here once more, but again, that would only exclude functions that are not quadratic anyway. The only real problem is having a negative discriminant. You cannot use factored form if $b^2 - 4ac < 0$.

Vertex to Factored

Q7 $r_1 = h - \sqrt{-\dfrac{k}{a}}$ $r_2 = h + \sqrt{-\dfrac{k}{a}}$

Q8 In this case, either k and a must have opposite signs, or k must be zero. Otherwise, r_1 and r_2 would both be undefined. This happens only when the function has no real roots.

Factored to Vertex

Q9 $h = \dfrac{r_1 + r_2}{2}$ $k = -\dfrac{a(r_1 - r_2)^2}{4}$

Students may give the formula for k in various unsimplified forms. They may even answer simply $k = f(h)$. If they have calculated h, then it is possible to do that in Sketchpad.

Changing Quadratic Function Forms

Below are three widely used forms of a quadratic function.

Standard: $ax^2 + bx + c$ Vertex: $a(x - h)^2 + k$ Factored: $a(x - r_1)(x - r_2)$

Which is better? That really depends on the situation. What information do you have? What do you need? It can even depend on your own personal style. Whatever the case, it is useful to be able to switch from one form to the other.

VERTEX TO STANDARD

Converting a function from vertex form to standard form is really just a matter of expanding the vertex expression.

$$ax^2 + bx + c = a(x - h)^2 + k$$

$$ax^2 + bx + c = a(x^2 - 2hx + h)^2 + k$$

$$ax^2 + bx + c = ax^2 - 2ahx + ah^2 + k$$

As you may already know, the parameter *a* has the same effect on all three quadratic function forms. Perhaps that's why we use the same variable name in all three versions.

If both sides express the same function, corresponding coefficients must be equal.

$$a = a \qquad b = -2ah \qquad c = ah^2 + k$$

1. Open **Changing Forms.gsp.** The "Vertex To Standard" page already has a quadratic function in vertex form along with its graph. Change the parameters to verify that everything works as you would expect.

To use the Calculator, choose **Number | Calculate.**

2. Use Sketchpad's Calculator to compute $-2ah$, and change the label of the result to b. Use the Calculator again to compute $ah^2 + k$, and change this label to c.

You will not be able to change the values of *b* or *c* directly. The parameters *a*, *h*, and *k* are still controlling everything.

3. Press the *Hide Vertex Form Graph* button. Choose **Graph | Plot New Function.** Enter the function in standard form: $g(x) = ax^2 + bx + c.$

$f(x) = a \cdot (x - h)^2 + k$

$y = 0.60[x - (2.70)]^2 + (-3.10)$

$g(x) = a \cdot x^2 + b \cdot x + c$

$y = 0.60x^2 + (-3.24)x + (1.27)$

By using the Hide/Show button, you can compare the two graphs to verify that the two function forms are equivalent. If the graphs do not overlay precisely, go back and check your calculations.

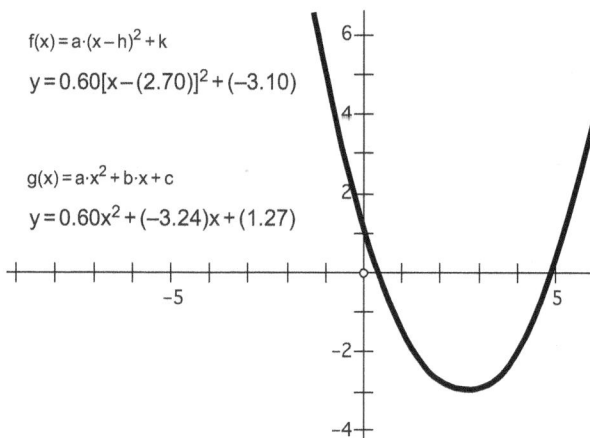

For more information on using Hot Text, press the *Hot Text Help* button.

4. To show the equation in standard form, you'll modify the red text, $y = a \cdot x^2 + b \cdot x + c$. Double-click the text to edit it, select the letter a, and click parameter a in the sketch. The parameter's value replaces the selected letter. Then select the red $+ b$ and hold the Shift key while clicking your calculated value of b. From the menu that appears, choose **Value as Addend.** (This command inserts the value preceded by the appropriate sign.) Similarly, replace the red $+ c$ with the appropriate calculated value as an addend.

Q1 What is $2(x - 8)^2 + 5$ in standard form?

STANDARD TO VERTEX

Converting standard form to vertex form is a bit more involved. Naturally, a is still a. One way to find b and c is to make use of these two equations from the previous section:

$$b = -2ah \qquad c = ah^2 + k$$

Use the first equation to write h in terms of a and b. Substitute the result into the second equation and write k in terms of a, b, and c.

Q2 What are h and k in terms of a, b, and c?

5. Open the "Standard To Vertex" page. Using the techniques from the previous section, construct a presentation for converting standard form to vertex form. Here is a summary of the steps:

- Create calculations for h and k.

- Create and plot a new function in vertex form using a, h, and k. Verify that the graph of the new function exactly overlays the original graph.

- Using the blue text and the calculations, insert the appropriate values in place of a, h, and k, and present the function form. (Because h is subtracted from x, you cannot insert it as an addend. Instead, calculate $-h$ and insert that calculation as an addend.)

Q3 Change parameter a to zero. The vertex form graph should disappear entirely. Does this mean that you can express certain quadratic functions in standard form but not vertex form? Explain.

OTHER CONVERSIONS

There are still four function conversions left. Open the document pages in order and use the summary in step 5 as a guide.

Factored to Standard

Moving from factored form to standard form works the same way as the conversion you did in the first section. Expand the expression.

Q4 How did you define b and c?

Standard to Factored

Remember that in factored form, $a(x - r_1)(x - r_2)$, the parameters r_1 and r_2 are the roots of the function. Use the quadratic formula to find the roots in terms of a, b, and c.

Q5 How did you define r_1 and r_2?

Q6 Under certain conditions, you cannot convert the function from standard form to factored form. What are those conditions? Explain.

Vertex to Factored

As in the previous case, the parameters r_1 and r_2 are the roots. You can find the roots by setting the function equal to zero and solving for x. Find the two solutions to this equation:

$$a(x - h)^2 + k = 0$$

Q7 How did you define r_1 and r_2?

Q8 Under what conditions is it not possible to convert a vertex form quadratic function into factored form?

Factored to Vertex

There is a geometric shortcut for this conversion. The coordinates of the vertex are (h, k) and the x-intercepts are r_1 and r_2. The vertex is on the axis of symmetry of the parabola. The x-intercept points are reflections of each other across the axis of symmetry. Therefore, the x-coordinate of the vertex must be the mean of the roots. That gives you h. The parameter k is $f(h)$.

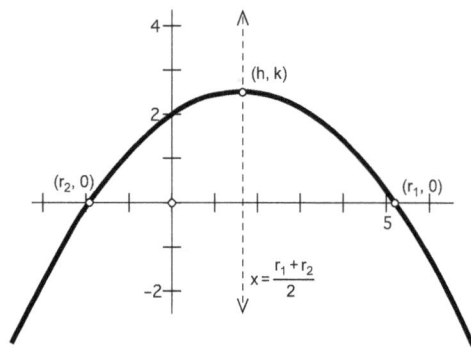

Q9 How did you define h and k?

3

Geometric Properties (Algebra)

The Pythagorean Theorem

COMPARING SQUARES

Q1 $a = 2, b = 3$, and $a^2 + b^2 = 2^2 + 3^2 = 13$

Q2 The area of a square is the square of the length of a side.

Q3 The area of square $CFGH$ is $5^2 = 25$.

Q4 Each triangle has an area of 3. Their combined area is 12.

Q5 If you remove the four triangles from the circumscribed square, what remains is the hypotenuse square, having area c^2.

$$c^2 = 25 - 12 = 13$$

This matches $a^2 + b^2$ from Q1, so $a^2 + b^2 = c^2$. This supports the theorem.

Q6 When $a = 4$ and $b = 7$, $a^2 + b^2 = 65$.

The area of the circumscribed square is $11^2 = 121$.

Each triangle has area 14, so their combined area is 56.

The area of the hypotenuse square is $121 - 56 = 65$, so again, $a^2 + b^2 = c^2$.

THE DISTANCE FORMULA

Q7 The coordinates of point C are (x_B, y_A).

Q8 $a = x_B - x_A$ and $b = y_B - y_A$

Q9 When B is to the left of A, the calculation for side b is negative. It has the correct magnitude, but the sign is negative. In the distance formula, the calculation is squared. The end result will be correct because $(-b)^2 = b^2$. The same applies to the calculation for a.

Q10 The calculation will match the measurement no matter what the positions of A and B are. This is one math formula with no special cases or exceptions.

Q11 $d = \sqrt{(x_B - x_A)^2 + (y_B - y_A)^2}$

WHOLE-CLASS PRESENTATION

Comparing Squares

1. Open **Pythagorean Theorem Present.gsp.** Drag points A and B around, and let the class see that the triangle vertices always fall on grid intersections.

Q1 Ask someone to explain what the Pythagorean theorem means as applied to this triangle. [The legs of the triangle are a and b, and the hypotenuse is c, so $a^2 + b^2 = c^2$.] Press the *Show Pythagorean Theorem* button.

2. Use small numbers the first time. Drag the vertices into a position such that $a = 3$ and $b = 2$.

3. Press the *Show Leg Squares* and *Show Hypotenuse Square* buttons.

Q2 Ask what the square areas represent. [They represent the squares of the three triangle sides.]

Q3 Show that you can substitute $a^2 = 4$ and $b^2 = 9$ into the equation. Challenge the class to find the area of the hypotenuse square without using the Pythagorean theorem.

4. Press the *Show Circumscribed Square* button.

Q4 Ask for the area of the large square [25].

Q5 Ask for the areas of the four triangles. [Each of them has an area of 3.]

Q6 Now ask again for the area of the hypotenuse square. Students should see that you can get this by subtracting the four triangle areas from the circumscribed square area: $25 - 4(3) = 13$

5. Make this substitution in the equation, showing that the Pythagorean theorem works in this one case. Drag the vertex points and try at least one other case.

Distance Formula

6. Go to page 2. You will see points A and B on the coordinate grid. Tell the class that the objective is to calculate the coordinate distance between the points using only their coordinates.

7. Press the *Show Coordinates* button.

Q7 Press the *Show Triangle* button. Ask the class for the coordinates of point C. [The coordinates are (x_B, y_A).]

Q8 Since the legs of the right triangle are vertical and horizontal, it is a simple matter to find their lengths by subtracting coordinates. Ask for the formulas: $a = y_B - y_A$, $b = x_B - x_A$.

8. Press the *Show Calculations for a and b* button.

Q9 At this point, you have the lengths of the legs of the right triangle. You can use the Pythagorean theorem to compute distance *d:*

$$d = \sqrt{(x_B - x_A)^2 + (y_B - y_A)^2}$$

9. Press the *Show Calculation for d* button.

10. Select points *A* and *B.* Choose **Measure | Coordinate Distance.** Compare this distance to the one you calculated. You can explain to the class that this measurement is actually found using the same formula.

Q10 Drag points *A* and *B* around the screen to confirm that the calculation and the distance always concur. Depending on the positions of the points, either of the calculations for *a* and *b* can be negative. That cannot be right, since these numbers represent segment lengths. Ask for a good explanation of why the formula works even when these signs are wrong. [It is because the formula squares both of these calculations, making their signs irrelevant.]

The Pythagorean Theorem

The Pythagorean theorem says that the sum of the squares of the lengths of the legs of a right triangle is equal to the square of the length of the hypotenuse. For the right triangle shown here, you can write the Pythagorean theorem as an equation.

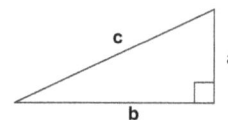

$$a^2 + b^2 = c^2$$

To prove the theorem, you must show that it is true for all right triangles. In this activity you will demonstrate the theorem for several specific cases.

COMPARING SQUARES

1. Open **Pythagorean Theorem.gsp.** You see right triangle *ABC* on a square grid.

Don't use **Measure | Distance** or **Measure | Area** in this activity, because you do not know the scale of the grid.

Q1 How long are *a* and *b*? What is the sum of the squares of the lengths of the legs ($a^2 + b^2$)? Don't measure—just count grid squares.

2. Press the *Show Leg Squares* button.

Q2 Explain why the areas of these squares are a^2 and b^2.

3. Press the *Show Hypotenuse Square* button.

This square, *ABDE*, has an area of c^2. Unfortunately, it is not aligned with the grid, so finding its area is more difficult.

4. Press the *Show Circumscribed Square* button.

Q3 This square, *CFGH*, fits around the hypotenuse square. What is its area?

Q4 There are four right triangles that also fit into the big square: $\triangle ABC$, $\triangle BDF$, $\triangle DEG$, and $\triangle EAH$. What is the area of each triangle? What is the sum of the triangle areas?

Q5 Use your answers to Q3 and Q4 to find the area of the hypotenuse square, *ABDE*. Does this support the Pythagorean theorem?

5. Drag the vertices of the triangle. You can change its dimensions, but it will always be a right triangle.

Q6 Change the leg dimensions to $a = 4$, $b = 7$. Using the same procedure as above, show that this triangle supports the Pythagorean theorem too.

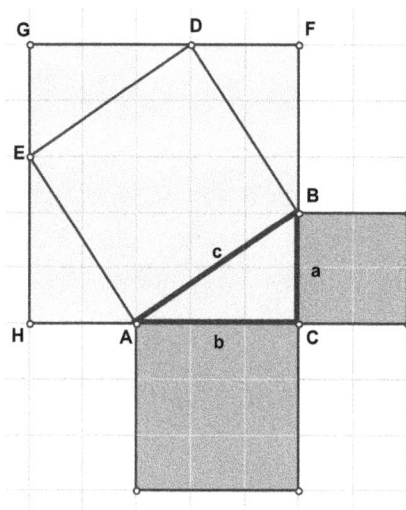

THE DISTANCE FORMULA

In the steps below you will use the Pythagorean theorem to find the distance between two specific points from their coordinates. You will finish by writing a general formula for the distance between any two points on the coordinate plane.

6. Go to page 2. This page contains points A and B on the coordinate grid.

7. Press the *Show Triangle* button. This shows you right triangle *ABC*, like the triangle in the previous section but with the hypotenuse labeled *d* for "distance."

Triangle *ABC* is a right triangle, so the Pythagorean theorem should apply.

$$d^2 = a^2 + b^2$$

$$d = \sqrt{a^2 + b^2}$$

8. Select points *A* and *B*. Choose **Measure | Abscissae (x)**. Select *A* and *B* again, and choose **Measure | Ordinates (y)**.

Q7 You don't have to measure the coordinates of point *C* separately. What are the coordinates of C in terms of x_A, x_B, y_A, and y_B?

Q8 In terms of the coordinates, what are the lengths of sides *a* and *b*?

To enter a coordinate into the Calculator, click the coordinate measurement in the sketch.

9. Choose **Number | Calculate**. Calculate sides *a* and *b* from the coordinates.

10. Choose **Number | Calculate**. Using your calculations for *a* and *b*, use the Pythagorean theorem to calculate distance *d*.

Q9 Drag point *B* so that it is left of *A*. What happens to the calculated distance *b*? Explain why this does not affect the final distance calculation.

11. Select *A* and *B*. Choose **Measure | Coordinate Distance**. Compare this distance to the one you calculated.

Q10 Drag points *A* and *B* to different locations on the coordinate plane. Does your calculated distance always match the measurement from step 11? Are there any special conditions under which the distance formula does not apply?

The label of your distance calculation should match the distance formula.

Q11 Write down a formula for the distance *d* in terms of the coordinates of *A* and *B*. This is the *distance formula*.

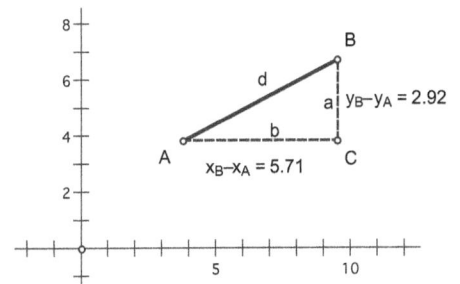

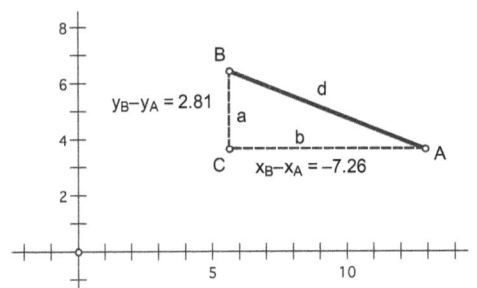

SKETCH AND INVESTIGATE

Q1 When the angles change, the ratios among the triangle's side lengths also change.

Q2 When the triangle changes scale without changing shape, the ratios among the triangle's sides don't change. The different triangles produced by dragging A or B are similar to each other, so their sides remain proportional.

Q3

$$\text{sine of } \angle A = \frac{\text{length of leg opposite } \angle A}{\text{length of hypotenuse}}$$

$$\text{cosine of } \angle A = \frac{\text{length of leg adjacent to } \angle A}{\text{length of hypotenuse}}$$

$$\text{tangent of } \angle A = \frac{\text{length of leg opposite } \angle A}{\text{length of leg adjacent to } \angle A}$$

Q4 $\sin 30° = 0.500$, $\cos 30° = 0.866$, $\tan 30° = 0.577$

Q5 The measure of $\angle C$ will vary depending on the measure of $\angle A$, but the two will always sum to 90°. The sine of $\angle C$ equals the cosine of $\angle A$. This is because the side opposite $\angle C$, side AB, is the side adjacent to $\angle A$.

Q6 a. The sine of an angle of 0° has value 0. This is the smallest value possible for the sine of an angle in a right triangle. (The sine of angles with negative values can get as small as -1. But negative angles cannot exist in a right triangle.)

b. The greatest possible value for the sine of an angle is 1. The angle with this sine is 90°. Although students won't be able to make $\angle A$ 90° in the sketch, they can see the sine approaching 1 as $\angle A$ approaches 90°.

c. If $\angle BAC = 90°$, the triangle has two right angles, which is impossible.

d. If you drag point C up as much as possible, the measure of $\angle A$ approaches 90°. At the same time, the length of the side opposite $\angle A$ gets larger and larger, while the length of the side adjacent to $\angle A$ stays the same. Since the tangent is the ratio of opposite/adjacent, the tangent ratio gets extremely large, approaching infinity as $\angle A$ approaches 90°.

e. The tangent equals 1 when the ratio of opposite/adjacent equals 1. This is true when the triangle is isosceles, which happens when $\angle A$ measures 45°.

f. The cosine and sine are equal when the triangle is isosceles. This means $\angle A$ measures 45°. The two ratios are equal because the sides opposite and adjacent to $\angle A$ are the same length.

g. $\sin x = \cos(90 - x)$

Trigonometric Ratios

Right-triangle trigonometry builds on similar-triangle concepts to give you more ways to find unknown measures in triangles. In this activity you'll learn about trigonometric ratios and how you can use them.

SKETCH AND INVESTIGATE

In steps 1–5, you'll construct a right triangle.

1. Construct $\overline{AB}$.

2. Construct a line through point B perpendicular to $\overline{AB}$.

Select point B and $\overline{AB}$; then, in the Construct menu, choose **Perpendicular Line.**

3. Construct $\overline{AC}$, where point C is a point on the perpendicular line.

$$m\angle CAB = 31°$$

$$\frac{m \text{ opposite}}{m \text{ hypotenuse}} = 0.51$$

$$\frac{m \text{ adjacent}}{m \text{ hypotenuse}} = 0.86$$

$$\frac{m \text{ opposite}}{m \text{ adjacent}} = 0.60$$

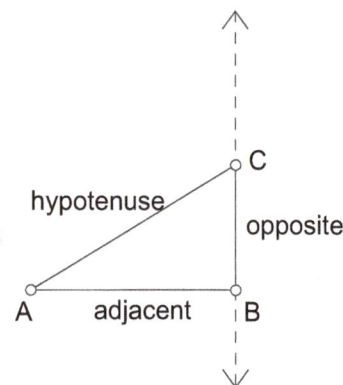

4. Hide the line.

Using the **Text** tool, click a segment to show its label. Double-click the label to change it.

5. Construct $\overline{BC}$ to finish the right triangle.

6. Show the three segments' labels and change the labels to match the figure above right.

Select, in order, points C, A, and B. Then, in the Measure menu, choose **Angle.**

7. Measure $\angle CAB$.

8. Measure the ratios *opposite/hypotenuse*, *adjacent/hypotenuse*, and *opposite/adjacent*.

For each ratio, select the two segments in order. Then, in the Measure menu, choose **Ratio.**

Q1 Drag point C to change the angles. When the angles change, do the ratios also change?

Q2 Drag point A or point B to scale the triangle. What do you notice about the ratios when the angles don't change? Explain why you think this happens.

Your observations in Q2 give you a useful fact about right triangles. For any right triangle with a given acute angle, each ratio of side lengths has a given value, regardless of the size of the triangle. The three ratios you measured are called *sine*, *cosine*, and *tangent*.

To measure an angle, press the **Marker** tool on the vertex and drag into the angle. Then select the angle marker and choose **Measure | Angle.** To measure a distance, select the two points and choose **Measure | Distance.**

9. The sine, cosine, and tangent functions can be found on all scientific calculators, commonly abbreviated as sin, cos, and tan. Use Sketchpad's Calculator to calculate the sine, cosine, and tangent of $\angle CAB$. Match these calculations with the ratios they are equal to.

Q3 Complete the ratios for cosine and tangent.

$$\text{sine } \angle A = \frac{\text{length of log opposite } \angle A}{\text{length of hypotenuse}}$$

$$\text{cosine } \angle A = \underline{\hspace{3cm}}$$

$$\text{tangent } \angle A = \underline{\hspace{3cm}}$$

Q4 Drag point C so that $\angle A$ measures as close to 30° as you can get it. Write approximate values for the sine, cosine, and tangent of 30° below. Use the definitions in Q3 and refer to the calculations in your sketch to find these values.

$\sin 30° = \underline{\hspace{2cm}}$ $\cos 30° = \underline{\hspace{2cm}}$ $\tan 30° = \underline{\hspace{2cm}}$

Q5 Without measuring, figure out the measure of $\angle C$ and write down that number. Calculate the sine of that angle measure. The sine of $\angle C$ should be close to one of the trigonometric ratios for $\angle A$. Which one? Explain why this is so.

Q6 Drag point C and answer the following questions.

a. What's the smallest possible value for the sine of an angle in a right triangle? What angle has this value?

Hint: Make $\overline{AB}$ short so that you can drag point C up farther.

b. What's the greatest possible value for the sine of an angle in a right triangle? What angle has this value?

c. Why can't you make m$\angle CAB$ exactly equal to 90°?

d. Even though you can't make m$\angle CAB$ exactly equal to 90°, what do you think is the value of tan 90°? Explain.

e. For what angle is the tangent equal to 1? Why?

f. For what angle are the sine and cosine equal? Why?

g. Suppose an angle has measure x. Complete this equation:

$$\sin x = \cos \underline{\hspace{3cm}}$$

SKETCH AND INVESTIGATE

Q1 The angle measurement approaches 0° and should eventually disappear when the slopes are equal.

Q2 The vertex E of $\angle AEC$ is the point of intersection of the lines. Lines with equal slope do not intersect, so the intersection point disappears when the slopes are the same; therefore, the angle also disappears.

Q3 If two lines have equal slopes, then the lines are parallel.

Q4 The product of the slopes of perpendicular lines is always -1 (as long as one of the lines is not horizontal).

Q5 If two lines are perpendicular, one of the slopes must be positive and the other negative. The product of a positive number and a negative number is always a negative number.

Q6 The slope of a vertical line is undefined, so the product of an undefined quantity with any other number is also undefined.

EXPLORE MORE

9. Select the line and choose **Equation** in the Measure menu. The coefficient of the x term in the equation equals the slope of the line (unless the line is vertical and has an undefined slope). For further development of this topic see the activity Different Slopes: The Slope of a Line.

10. Students should locate the second point by using the same rise and run as the two points on the first line.

11. If students construct parallel lines, the slopes will always be exactly equal. The product of the slopes of constructed perpendicular lines will always be exactly -1.

Slopes of Parallel and Perpendicular Lines

In this investigation you'll learn how you can use slope to tell whether lines are parallel or perpendicular.

SKETCH AND INVESTIGATE

Choose Preferences from the Edit menu and go to the Units panel.

1. In Preferences, set Angle Precision to **tenths** and precision of **Others** to **hundredths.**

2. Construct $\overleftrightarrow{AB}$ and $\overleftrightarrow{CD}$ and their point of intersection, *E*.

$m\angle AEC = 38.3°$

Slope $\overleftrightarrow{AB} = 0.56$

Slope $\overleftrightarrow{CD} = -0.16$

Select, in order, points *A*, *E*, and *C*; then, in the Measure menu, choose **Angle.**

3. Measure $\angle AEC$.

4. Measure the slopes of $\overleftrightarrow{AB}$ and $\overleftrightarrow{CD}$.

While holding down the Shift key, choose **Hide Coordinate System** from the Graph menu.

5. Hide the coordinate system that appeared when measuring the slopes.

6. Drag point *A* and observe the slope measures.

Q1 Make the slopes as close to equal as you can. What do you observe about the measure of the angle between the lines?

Q2 If you get the slopes close enough to equal, the angle measure will actually disappear. Why do you think that happens? (*Hint:* The vertex of this angle is the point of intersection of the two lines.)

Q3 Write a conjecture about lines with equal slopes.

Choose **Calculate** from the Number menu to open the Calculator. Click a measurement to enter it into a calculation.

7. Calculate the product of the slopes of $\overleftrightarrow{AB}$ and $\overleftrightarrow{CD}$.

8. Make sure that neither line is horizontal. Drag points to make $m\angle AEC$ as close to 90° as you can.

Q4 What is the product of the slopes of perpendicular lines? _____

Q5 Why is this product always negative?

Q6 The product of the slopes of two lines is undefined if one of the lines is vertical. Why?

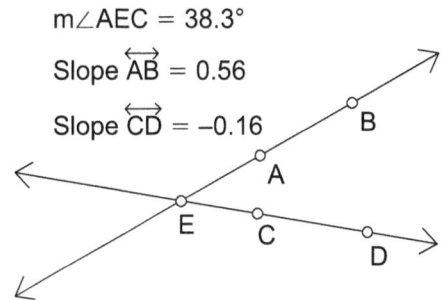

EXPLORE MORE

9. In the same sketch, mesure the equations of the two lines. Where does the slope of a line appear in its equation?

It may help to choose
Snap Points from the
Graph menu.

10. In a new sketch, show the coordinate grid. Scale the grid, if necessary, so that grid points are about 1/2 in. (or 1 cm) apart. Hide the axes. Draw a line and a point not on the line. Now construct a second point not on the line, located so that when you draw a second line through these points it will be parallel to the first line. Explain how you located the second point.

11. Confirm your parallel-line slope conjecture by constructing a line and a point not on the line. Through the point not on the line, construct a parallel line. Measure the slopes of the two lines. Drag different points and observe the slope measurements. Do a similar investigation for perpendicular lines. Explain what you did and what your investigations demonstrate.

The Distributive Property:
A Painting Dilemma

To use this activity with measurements in meters rather than feet, use pages "1(m)" and "2(m)", and use the **Scaled Length (m)** tool.

SKETCH AND INVESTIGATE

Q1 The result for Cari's method is

$$(25 \text{ ft})(80 \text{ ft}) + (25 \text{ ft})(100 \text{ ft})$$
$$= 2000 \text{ sq ft} + 2500 \text{ sq ft}$$
$$= 4500 \text{ sq ft}$$

Q2 The result for Zeeba's method is

$$(25 \text{ ft})(80 \text{ ft} + 100 \text{ ft})$$
$$= (25 \text{ ft})(180 \text{ ft})$$
$$= 4500 \text{ sq ft}$$

Q3 $ab + ac = a(b + c)$

Q4 Both methods still give the same result. For these particular numbers, the result is 3500 sq ft.

Q5 In the case of three walls, the factor a would be multiplied by the widths of all three walls:

$$ab + ab + ac = a(b + b + c) = a(2b + c)$$

THREE WALLS

Q6 Both methods do give the same result: 6500 sq ft.

Q7 See Q5.

Q8 Answers will vary. In the perspective view the scale is variable. This is because objects are foreshortened, and some objects are closer to the viewer than others. Moreover, a rectangle generally does not even appear as a rectangle.

WHOLE-CLASS PRESENTATION

Use **Distributive Painting Present.gsp** in conjunction with the Presenter Notes to present this activity to the whole class.

1. Open **Distributive Painting Present.gsp,** and explain the problem to your class based on the description at the beginning of the student activity sheet.

2. Press the *Cari's Solution* button. Explain that this is the configuration on which Cari decided to base her calculation.

3. Explain that the first part of Cari's method is to calculate the area of the rectangle on the left. Choose **Number | Calculate** to show the Calculator. Click the 25 ft measurement in the sketch, the multiplication sign on the keypad, and the 80 ft measurement in the sketch. Click **OK** to finish the calculation.

4. Use a similar calculation to compute the area of the rectangle on the right. Then use the Calculator one more time to find the sum.

Q1 Ask, "How would you express this calculation using the variables *a*, *b*, and *c*?" (The label of the calculation gives the answer away. Congratulate students for noticing this.)

Now calculate the area using Zeeba's method.

5. Press the *Reset* button to return the walls to their original arrangement. Press the *Zeeba's Solution* button.

6. Use the Calculator to multiply the height (25 ft) by the length (180 ft). Be sure to click the measurements in the sketch; don't type the numbers on the keypad.

Q2 Ask, "How would you express this calculation using the variables *a*, *b*, and *c*?" (The label of the calculation again gives the answer away. Don't congratulate the students this time.)

Q3 Ask, "Are the results equal?"

Q4 Ask, "Will the results always be equal, even if the dimensions are different?"

Even if the entire class says the results will always be equal, tell them that it's too easy to be fooled in problems like this and that it's worth trying different measurements.

7. Press the *Show Dimensions* button, double-click the *height* measurement, and change it to 20. Double-click the *height* measurement and change it to 60. Press *Reset.*

8. Press the *Cari's Solution* button to show Cari's result, and then the *Zeeba's Solution* button to show Zeeba's.

Q5 Ask students to summarize the principle of algebra that they've found by investigating this problem.

9. If there's time, explore page 2 (with three walls).

The Distributive Property: A Painting Dilemma

The school activities committee is preparing to paint two gymnasium walls. Both walls are 25 ft high. The first wall is 80 ft wide, and the other is 100 ft wide.

Cari and Zeeba are on the committee, and they volunteered to calculate how many cans of paint to order. To figure this out, they need to find the total area of the walls. But Cari and Zeeba disagree about how to do the calculation.

Cari wants to calculate the area of each wall separately and then add them together to get the total painted area:

$$(25 \text{ ft})(80 \text{ ft}) + (25 \text{ ft})(100 \text{ ft})$$

Zeeba wants to calculate the total area by first adding up the widths of the walls to get the total width and then multiplying the height by this total:

$$(25 \text{ ft})(80 \text{ ft} + 100 \text{ ft})$$

SKETCH AND INVESTIGATE

1. Open **Distributive Painting.gsp.** This is a perspective view of the two walls as seen from inside the gym.

First calculate the area using Cari's method.

2. Press the *Cari's Solution* button.

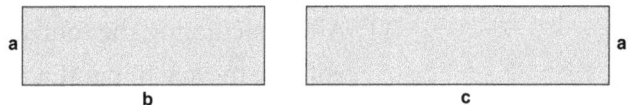

3. To measure the walls, press and hold the **Custom** tool icon and choose **Scaled Length.** Click this tool on line segments *a*, *b*, and *c*.

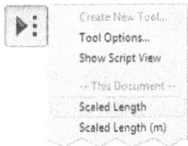

4. To calculate the area using Cari's method, first choose **Number | Calculate** and find the area of the rectangle on the left. To enter the width and height of this wall into the Calculator, click the measurements in the sketch.

5. Calculate the area of the rectangle on the right. Then finish calculating by Cari's method by using the Calculator one more time to find the sum of the two areas.

Q1 What is the result using Cari's method?

Now calculate the area using Zeeba's method.

6. Press the *Reset* button to return the walls to their original arrangement. Press the *Zeeba's Solution* button.

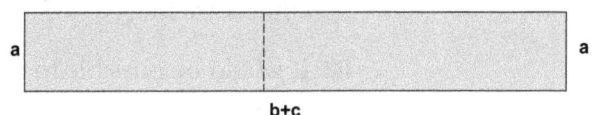

7. Measure the line segment labeled *b* + *c*.

If you prefer, you can do all three calculations in one step.

8. Calculate the area using Zeeba's method, by using Sketchpad's Calculator to calculate the total area of this single rectangle.

Q2 What is the result using Zeeba's method?

Q3 Write your expressions as an equation using the variables *a*, *b*, and *c*. This equation is a symbolic statement of the *distributive property of multiplication over addition.*

You've tried your result for only one set of measurements. Jason claims that it might work for this one case, but it won't work for others.

9. To test Jason's claim, press the *Show Dimensions* button, change the length from 100 ft to 60 ft, and press *Reset.* Then press the buttons to try both Cari's solution and Zeeba's solution.

> To change the length, double-click it and type a new value.

Q4 Do both methods still give the same result?

Q5 Make a conjecture. How would the distributive property apply if there were three walls, two with width *b* and one with width *c*?

THREE WALLS

10. Go to page 2 of the document.

11. After calculating the total area of the walls, the students find that they have enough money to paint a third wall in the gym. Repeat the previous steps and calculate the painted area using both Cari's and Zeeba's methods.

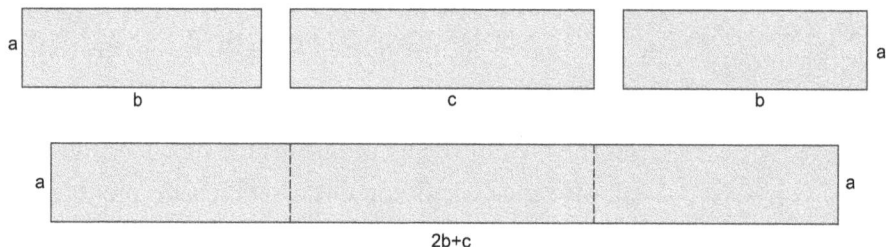

Q6 Do both methods give the same result? What is the combined area of the three walls?

Q7 Write your expression as an equation. Compare the equation with your conjecture in Q5.

Q8 It would be possible to measure the line segments while the sketch is showing a perspective view, but the results would not be very useful. Explain why this is true.

A Rectangle with Maximum Area

SKETCH AND INVESTIGATE

Q1 As students drag point *C*, they should notice that the area of the rectangle changes but its perimeter remains constant. Because *CB* and *CD* are radii of the same circle, the sum of two sides of the rectangle, *AC* + *CD*, is equal to *AB*. Thus, *AB* is half the perimeter of the rectangle. As long as this length is kept constant, the perimeter of the rectangle will be constant.

Q2 A square is the rectangle with the greatest area for a given perimeter.

Q3 The coordinates of the high point of the graph show the side length and area of the maximum-area rectangle. The side length at this point verifies that the rectangle with the maximum area is a square.

Q4 The low points on the graph show where the area of the rectangle is zero. This happens when *AC* is zero and when *AC* = *AB*.

You might want to discuss with students why the locus graph of (side length, area) of a rectangle is a parabola.

EXPLORE MORE

16. Regular polygons have maximum area for a given perimeter. Polygons with more sides are more efficient. The circle is the closed planar figure that gives maximum area for a given perimeter.

17. The area of the rectangle can be represented by the equation $A = x[(1/2)P - x]$. The graph is a parabola with roots 0 and $(1/2)P$. So the *x*-value of the maximum point is $(1/4)P$. Since the side length of the maximum area rectangle is 1/4 the rectangle's perimeter, the rectangle must be a square.

A Rectangle with Maximum Area

Suppose you had a certain amount of fence and you wanted to use it to enclose the biggest possible rectangular field. What rectangle shape would you choose? In other words, what type of rectangle has the most area for a given perimeter? You'll discover the answer in this investigation. Or, if you have a hunch already, this investigation will help confirm your hunch and give you more insight into it.

SKETCH AND INVESTIGATE

1. Construct $\overline{AB}$.

Select $\overline{AB}$, point A, and point C. Then, in the Construct menu, choose **Perpendicular Line.**

2. Construct $\overline{AC}$ on $\overline{AB}$.

3. Construct lines perpendicular to $\overline{AB}$ through points A and C.

Be sure to release the pointer—or click a second time—with the pointer over point B.

4. Construct circle CB.

5. Construct point D where this circle intersects the perpendicular line.

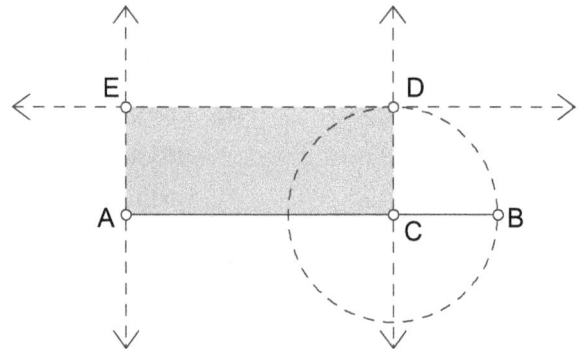

6. Construct a line through point D, parallel to $\overline{AB}$.

7. Construct point E, the fourth vertex of rectangle $ACDE$.

Select the vertices of the rectangle in consecutive order. Then, in the Construct menu, choose **Quadrilateral Interior.**

8. Construct interior $ACDE$.

9. Measure the area and perimeter of this polygon.

10. Drag point C back and forth and observe how this affects the area and perimeter of the rectangle.

Select point A and point C. Then, in the Measure menu, choose **Distance.** Repeat to measure AE.

11. Measure AC and AE.

Q1 Without measuring, state how AB is related to the perimeter of the rectangle. Explain why this rectangle has a fixed perimeter.

Q2 As you drag point C, observe what rectangular shape gives the greatest area. What shape do you think that is?

In steps 12–14, you'll explore this relationship graphically.

Select, in order, measurements AC and Area $ACDE$. Then choose **Plot as (x, y)** from the Graph menu. If you can't see the plotted point, drag the unit point at (1, 0) to scale the axes.

12. Plot the measurements for the length of $\overline{AC}$ and the area of $ACDE$ as (x, y). You should get axes and a plotted point H, as shown in the following step.

13. Drag point C to see the plotted point move to correspond to different side lengths and areas.

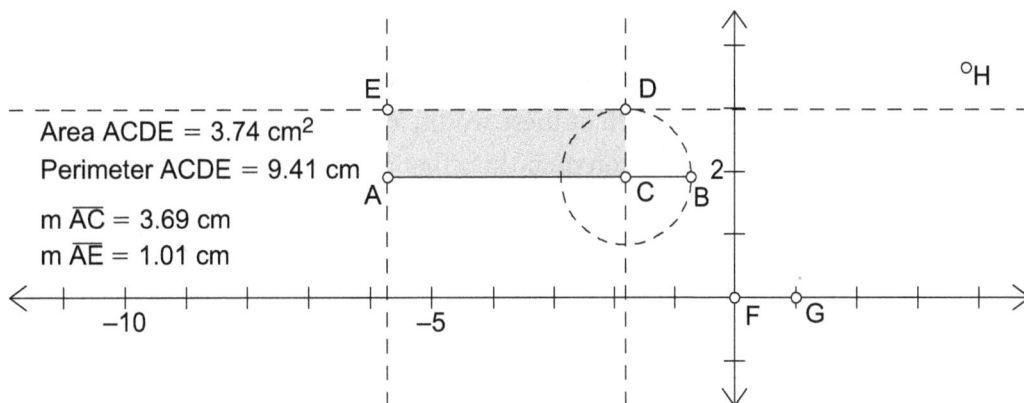

Area ACDE = 3.74 cm²
Perimeter ACDE = 9.41 cm

m $\overline{AC}$ = 3.69 cm
m $\overline{AE}$ = 1.01 cm

Select point *H* and point *C;* then, in the Construct menu, choose **Locus.**

You may wish to select point *H* and measure its coordinates.

14. To see a graph of all possible areas for this rectangle, construct the locus of plotted point H as defined by point C. It should now be easy to position point C so that point H is at a maximum value for the area of the rectangle.

Q3 Explain what the coordinates of the high point on the graph are and how they are related to the side lengths and area of the rectangle.

15. Drag point C so that point H moves back and forth between the two low points on the graph.

Q4 Explain what the coordinates of the two low points on the graph are and how they are related to the side lengths and area of the rectangle.

EXPLORE MORE

16. Investigate area/perimeter relationships in other polygons. Make a conjecture about what kinds of polygons yield the greatest area for a given perimeter.

17. What's the equation for the graph you made? Let AC be x and let AB be $(1/2)P$, where P stands for the perimeter (a constant). Write an equation for area, A, in terms of x and P. What value for x (in terms of P) gives a maximum value for A?

Direct Variation

In this activity students move from looking at properties of lines (in particular, slope) to generating linear relationships. Before starting, you might have a discussion about quantities that grow in different ways (proportionally, exponentially, and inversely) in relation to each other—and how each of these would look on a graph. The activity Inverse Variation focuses on quantities that are inversely proportional.

This activity focuses on lines of the form $y = mx$ (or $y = bx$), and another activity (The Slope-Intercept Form of a Line) focuses on lines of the form $y = mx + b$. It would be valuable to make the connection here between direct relationships and linear relationships (of which direct relationships are a subset).

SKETCH AND INVESTIGATE

1. If Sketchpad is set to its default Preference settings, points won't be labeled when they are created. Students can click points with the **Text** tool to label them. (Points will be labelled in alphabetical order.) To edit a label, double-click it with the **Text** tool.

Q1 Dragging B changes only the base and the area. Dragging C changes only the height and the area.

12. The sketch becomes cluttered at this point. Students may want to move the origin down near the bottom of the sketch window, hide the grid, and move the rectangle to a relatively clear area of the sketch.

Q2 It shows that as the height gets bigger, the area gets bigger; that as the height gets smaller, the area gets smaller; and that they grow or shrink proportionally to each other.

Q3 $A = base \cdot height$

Q4 $f(x) = base \cdot x$

Q5 It's the same. (To be more precise, it contains the path of the plotted point; the function exists in the first and fourth quadrants, but the plotted point is always in the first quadrant.)

Q6 The graph passes through the origin because the area of a rectangle with height 0 is 0—hence the point $(0, 0)$. And algebraically, when $x = 0$ in $f(x) = base \cdot x$, $f(0) = 0$—hence the point $(0, 0)$.

Q7 A rectangle can't have a negative height or area. The domain should be restricted to $x > 0$ (or possibly $x \geq 0$ if you consider 0 to be a possible height of a rectangle).

Q8 It means that as one quantity doubles, the other doubles; as one triples or halves, the other triples or halves. For example, the area of a rectangle with base 3 and height 4 is 12. If you double the height to 8 (and leave the base the same), the area also doubles to 24. The word "proportional" is used because the area and the height are in proportion ($12/4 = 24/8 = 3$). The base is the constant of proportionality.

Q9 This changes the slope of the line. Students may also notice that the plotted point moves vertically up and down (which makes sense because the x-value, which represents height, is not changing).

Q10 The length of the base is the slope of the graph. "Wide" rectangles (those with larger bases) will have steeper graphs. The reason is that every increase in height will add a lot to the area. "Skinny" rectangles (those with smaller bases) will have more gradual graphs. The reason is that similar increases in height will add much less to the area.

EXPLORE MORE

Q11 The point traces out a portion of a parabola. The trace is no longer linear, so this is *not* direct variation. The reason this happens is that we are now varying both the height and the base simultaneously, whereas before we were varying only the height, leaving the base constant. Variation in one dimension results in a linear graph, whereas variation in two dimensions results in a quadratic graph.

WHOLE-CLASS PRESENTATION

Use the sketch **Direct Variation Present.gsp** to explore with the class the relationships between measurements in a rectangle. The goal is to use this visual tool (and hopefully students' intuition) to think about why certain quantities are proportional and to connect this proportionality to direct variation via an equation and a graph.

Use page 1 of the sketch to answer Q1–Q10 together as a class. Use page 2 for Explore More (optional), which features a square instead of a rectangle. In this case, the area also increases when the side length increases, but the relationship is quadratic rather than linear.

Direct Variation

What happens to the area of a rectangle if you keep the length of the base constant while varying the height? (Try to answer this question before reading on.) What happens if you enlarge the entire rectangle? In this activity you will learn about direct variation and how it's represented algebraically and graphically.

SKETCH

Start by constructing a rectangle and its interior.

1. In a new sketch, construct a point and label it *A*.

To translate A, select it and choose **Transform | Translate.** *To construct the ray, select the two points and choose* **Construct | Ray.**

2. Translate point *A* by 1.0 cm at 0°. Construct a horizontal ray from *A* through the translated point.

3. Hide the translated point, construct a new point on the ray, and label it *B*.

4. Translate *B* by 1.0 cm at 90°. Construct a vertical ray from *B* through the translated point.

5. Hide the translated point, construct a new point on the ray, and label it *C*.

Select B and A in order and choose **Transform | Mark Vector.**

6. Mark the vector from point *B* to point *A*.

7. Translate point *C* by the marked vector. Label the translated point *D*.

8. Hide the rays and construct $\overline{AB}$, $\overline{BC}$, $\overline{CD}$, and $\overline{DA}$.

You should have a rectangle. Drag each of the four points to be sure it remains a rectangle and to see how dragging each point changes the rectangle.

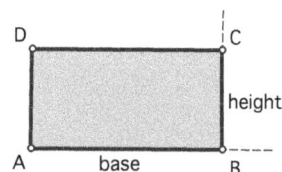

Select the four points in order and choose **Construct | Quadrilateral Interior.**

9. Construct polygon interior *ABCD*. Measure its area.

10. Measure the lengths of base $\overline{AB}$ and height $\overline{BC}$ by selecting them and choosing **Measure | Length.**

11. Click the **Text** tool on $\overline{AB}$ and $\overline{BC}$ to show their labels. Change $\overline{AB}$'s label to *base* and $\overline{BC}$'s label to *height* by double-clicking the **Text** tool on each label.

INVESTIGATE

Next you'll investigate how the area changes as you change the size of the rectangle.

Q1 Drag the points around. How do the measurements change when you drag *B*? How do the measurements change when you drag *C*?

If the sketch is too cluttered, choose **Graph | Hide Grid.**

12. A graph will help to show what's happening. Select in order the *height* and *Area ABCD* measurements, and choose **Graph | Plot as (x, y).**

You have just plotted a point whose x- and y-coordinates are the height and area of the rectangle, respectively. (If you can't see it yet, you will soon.)

To trace an object, select it and choose **Display | Trace Plotted Point.** To clear traces from the screen, choose **Display | Erase Traces.**

13. Drag point C closer to $\overline{AB}$ and observe the effect on the plotted point. Trace the plotted point so you can see what it does as you drag point C.

Q2 What does the path of the plotted point tell you about how height and area are related in a rectangle when the base is kept constant?

Q3 Write the formula for the area of a rectangle in terms of *base* and *height*.

Q4 Now write the same formula as a function using $f(x)$ for *area* and x for *height*.

To enter *base* in your formula, click its measurement in the sketch. To enter x, click the x key on the Calculator's keypad.

Q5 Plot your function from Q4 by choosing **Graph | Plot New Function.** How does the function plot relate to the path of the plotted point as you vary the height of the rectangle?

Q6 Why does it make sense that the graph passes through the origin?

Q7 The function is plotted as a line. But the traces cover only part of the line. Why? If you wanted the function plot to accurately represent the situation, what part of it would you cut off? (In other words, how would you restrict the *domain*?)

Q8 We say that a rectangle's area *varies directly with* (or *is directly proportional to*) its height when its base is held constant. Describe in your own words what you think this means.

To answer Q10, compare a rectangle that produces a steep graph to one that produces a gradual graph.

Q9 Erase the traces. Drag point B to change the length of the base of the rectangle. What effect does this have on the graph?

Q10 How is the length of the base related to the slope of the graph?

EXPLORE MORE

Q11 Go to page 2 of **Direct Variation.gsp.** This page shows square *ABCD*. Does the area of the square vary directly with the length of one side? Make a prediction and then drag the points to check your prediction. Describe the path of the plotted point. Is this direct variation? Why or why not?

4

Measurement, Data, and Probability (Algebra)

Some textbooks describe "cross-multiplying" of proportions, and some avoid the technique. Q11 asks students to justify this technique. This question is optional, and you can tell students to skip it if you prefer.

EXAMINE YOUR ASSUMPTIONS

The model used in this activity is one way to see equal ratios. The visual model is designed to illustrate repeating a rate—every 4 seconds, you will see 7 more noodles. The model also marks off, with the red segments, groups of 7 noodles and 4 seconds, so that you can see that you have made 8 copies of the original ratio. The discrete nature of the machine makes it easier for students to see these copies. If students are actually counting noodles and seconds, they must count the colored intervals between the vertical line segments, not the segments themselves.

Students can change the rate at which their machine works by changing the rate parameters at the top of the sketch: the number of noodles and the number of seconds.

Q1 If you are buying 3 boxes of detergent, and each box has the same number of ounces and costs the same, then you can repeat the ratio in this way. However, students may be familiar with buying in bulk for a lower cost.

Q2 The runner may be able to keep up that same speed while running 10 times as far, but then again, she may not.

Q3 The machine produces 7 noodles in one turn of the crank, which takes 4 seconds. The ratio is

$$\frac{7 \text{ noodles}}{4 \text{ seconds}}$$

Q4 After two turns of the crank, there are 14 noodles, and the machine has been running for 8 seconds.

Q5 There are 8 copies of 4 seconds in 32 seconds, so there are 8 copies of 7 noodles, making 56 noodles.

Q6 The pasta machine produces the same number of noodles every 4 seconds and doesn't slow down or speed up over time. Some students will argue that this means it runs at a constant rate. Other students may point out that the machine doesn't produce any finished noodles until the end of the 4 seconds, so it's not really running at a constant rate—and that at 2 seconds, it has not produced even a single finished noodle. Both arguments have merit, and this question can lead to a very enlightening class discussion on what is meant by *constant rate*, and on the difference between discrete behavior (exemplified by the pasta machine on page 1) and continuous behavior (exemplified by the machine on page 2).

RATES AND PROPORTIONS

No matter how a proportion is solved, students should have a sense of the meaning of their calculations. Solving by finding the number of copies as in Q7 focuses students on the idea that they are repeating a rate.

Q7 You must repeat the 4-second ratio 7.5 times. You can see this in the model by thinking of 4 as a unit; there are 7.5 groups of 4 seconds in 30 seconds. The red segments shown in the model may help. Since you have 7.5 groups of 4 seconds, you will also have 7.5 groups of 7 noodles, or 52.5 noodles.

Q8 Calculate 30 seconds divided by 4 seconds times 7 noodles. Dividing 30 by 4 means 7.5 groups of 4 seconds. Since the rate is constant, there will be 7.5 groups of 7 noodles. Multiply 7.5 by 7.

Q9 Divide to find the number of times the ratio will be repeated. Multiply this number by the other quantity you are repeating.

Q10 Since 105 noodles equals $2y$ noodles, the value of y is 52.5. The two rates (y noodles in 30 seconds and 7 noodles in 4 seconds) are equal, so repeating both will result in ratios that are still equal. Whichever equation you use to solve for y, the value of y is the same.

Q11 Take the present question as an example. You would multiply 30 by 7, and 4 by y. The meaning of these steps is clearer if you write

$$\frac{30(7 \text{ noodles})}{30(4 \text{ seconds})} = \frac{4(y \text{ noodles})}{4(30 \text{ seconds})}$$

This means that you have repeated the ratio on the left 30 times and the one on the right 4 times. Since each ratio gives the number of noodles made in 120 seconds, the numerators must be equal. Since $4y$ equals 210, y equals 52.5.

New York City Title I High School Activities with The Geometer's Sketchpad
© 2012 Key Curriculum Press

EXPLORE MORE

Q12 To calculate a unit rate equivalent to the given rate of 7 noodles in 4 seconds, divide both the numerator and denominator by 4. The unit rate is 1.75 noodles per second. Knowing how many noodles are made in a second allows you to multiply by 30 to find how many noodles were made in 30 seconds. As before, you are repeating a constant rate.

Q13 Divide to find a unit rate; multiply this unit rate by the number of units to which you are repeating the rate.

WHOLE-CLASS PRESENTATION

In this presentation students will observe a machine running at a certain rate and use a proportion to figure out how much it will produce during a certain period of time. Students will discuss what a constant rate means, why it's required to use a proportion, and how continuous and discrete processes differ.

Start with a Sketchpad model of a pasta machine producing noodles at a certain rate.

1. Open **Rates and Ratios.gsp.** Press the *Turn the Crank* button.

Q1 Ask, "How many noodles did the machine produce in one turn of the crank? How long did it take?" (Students can use the gold rectangles to count the noodles, and the blue ones to count the seconds.)

Q2 Ask students to write the ratio of noodles to seconds as a fraction.

Q3 Turn the crank again and ask students how many noodles there are now, and how many seconds the machine has been running.

2. Turn the crank 6 more times, so the machine has run for a total of 32 seconds.

Q4 Ask students to determine how many noodles the machine has made without counting them. Ask how they got their answers. Express the student explanations as proportions.

Discuss what a *constant rate* means.

Q6 Ask whether the pasta machine runs at a constant rate. This question does not have a clear-cut answer, and should lead to an interesting discussion about the meaning of constant rate and about discrete and continuous processes.

A slightly different machine can help students understand the idea of constant rate.

3. Go to page 2 and press the *Turn the Crank Once* button.

Q7 Ask students how this machine differs from the machine on page 1. Turn the crank again to help them observe the differences.

Q8 Ask students to figure out how many noodles this machine can produce in 30 seconds. They should do their calculation by setting up a proportion. Once they have their answers, check the answers by dragging point *time* to 30 seconds.

Have students summarize the results.

Q9 Ask students to describe the method they used to find the number of noodles so that another person could use the method to solve any proportion.

Q10 Ask students to explain why using proportions works only if the machine runs at a constant rate.

Rates and Ratios

You buy 52 ounces of detergent for 9 dollars. You run 100 meters in 15 seconds. You use 5 gallons of gas to drive 150 miles. Each of these pairs of numbers is an example of a *ratio*.

You can use ratios to answer questions like: If my pasta machine makes 7 noodles in 4 seconds, how many noodles will it make in 30 seconds? To answer such questions, you have to make some assumptions. In this activity, you will examine what those assumptions are and use them to answer questions like the ones posed here.

EXAMINE YOUR ASSUMPTIONS

This ratio is sometimes written as 52:9, and sometimes as $\frac{52}{9}$.

Suppose you can buy 52 ounces of detergent for 9 dollars. If you triple both the values in that ratio, you'll get 156 ounces for 27 dollars. The second ratio is considered equal to the first, because it consists of 3 copies of the original ratio.

Any pair of numbers, like the ones mentioned above, can be considered a ratio. Not all ratios, however, can be applied in the same way.

Q1 In your experience, if 52 ounces of detergent cost 9 dollars, will 156 ounces cost 27 dollars? Why or why not?

Q2 In your experience, if a person ran 100 meters in 15 seconds, will that person run 1000 meters in 150 seconds? Why or why not?

A *constant rate* is a rate that stays the same over time.

In both examples above, you can create a new ratio equal to the original ratio. Whether the new ratio is meaningful depends upon whether your ratio represents a *constant rate* (a rate that stays the same over time).

Imagine that each turn of the crank on your pasta machine takes a certain amount of time and produces a certain number of noodles.

1. Open **Rates and Ratios.gsp.**

2. To run the machine, press the *Turn the Crank* button.

Use the gold rectangles to count the noodles, and the blue ones to count the seconds.

Q3 How many noodles does the machine produce in one turn of the crank? How long does it take? Write the ratio of noodles to seconds as a fraction.

Q4 Turn the crank again. How many noodles do you have now? How many seconds has the machine been running?

Q5 Keep turning the crank until the machine has run for 32 seconds total. Without counting them, determine how many noodles the machine has made. How did you figure this out without counting?

Q6 Does the pasta machine run at a constant rate? Explain.

RATES AND PROPORTIONS

A *proportion* is a statement that two ratios are equal. You used the pasta machine to show the rate at which the machine runs in the form of two different ratios:

$$\frac{7 \text{ noodles}}{4 \text{ seconds}} = \frac{56 \text{ noodles}}{32 \text{ seconds}}$$

The second ratio extends the basic rate (7 noodles in 4 seconds) 8 times.

3. Go to page 2. This model of the pasta machine runs at the same rate but produces noodles continuously rather than in batches.

Since the machine runs at a constant rate, the rate for 4 seconds is the same as the rate for 30 seconds, and you can use the proportion

$$\frac{7 \text{ noodles}}{4 \text{ seconds}} = \frac{y \text{ noodles}}{30 \text{ seconds}}$$

to determine how many noodles the machine can make in 30 seconds.

Model this by dragging the time slider—but don't count your noodles!

Q7 How many times do you need to repeat the rate "7 noodles in 4 seconds" to reach 30 seconds? Explain how to use this value to determine the value of y.

Q8 Describe a series of calculations that produces the value of y using the quantities 7 noodles, 4 seconds, and 30 seconds. Explain why this method works.

Q9 Describe the method you used in the previous question so that another person could use the method to solve any proportion.

Q10 If you repeat the ratio "7 noodles in 4 seconds" 15 times, you get $15 \cdot 7$ noodles in $15 \cdot 4$ seconds. That's 105 noodles in 60 seconds. If you repeat the ratio "y noodles in 30 seconds" 2 times, that's $2y$ noodles in 60 seconds. What is the value of y? Should it be the same as the value you found in Q7? Why?

Q11 If you've heard about "cross-multiplying" as a way to solve a proportion, explain why it works using the idea of "repeating" both ratios in a proportion.

EXPLORE MORE

Q12 Reduce the number of seconds to exactly 1 by dragging point *time*. How many noodles does the machine make in 1 second? How did you calculate this? Does multiplying this value by 30 give you the number of noodles made in 30 seconds? Why?

Q13 Describe the method you used in the previous question so that another person could use the method to solve any proportion.

SKETCH AND INVESTIGATE

Q1 See A1 in the student section.

Q2 No, this is not possible. The middle half of the data must fall within the box, and the median must be somewhere in that middle half.

Q3 Yes. It will happen if the middle half of the data have the same value somewhere in the middle of the range.

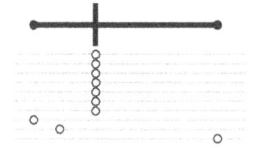

Imagine a math test with only two questions, one very easy and one very difficult. Nearly all of the students would score 1, but a few outliers could score 0 or 2.

Q4 Yes. It is similar to the distribution in Q1, except that there is both a lower limit and an upper limit.

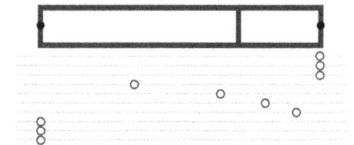

Show a photo of the cast of *Gilligan's Island* to a large group of people. Ask them to identify the characters. Many people have no knowledge of the show, and would score zero. Of the people who have seen the show, most would have no difficulty naming all seven.

Q5 Yes. This can be caused by a fairly close grouping with some extreme outliers.

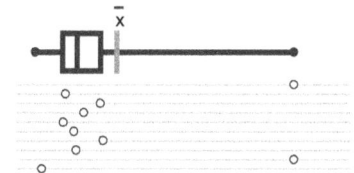

This situation has been modeled in real life in baseball labor disputes. The mean income of professional baseball players is very high. However, most professional players play in the minor leagues and earn a modest income. Relatively few outliers in the majors earn exorbitant salaries, raising the mean but having little effect on the median or the quartiles.

Q6 No, this is not possible. The mean must be within the range of the sample. You can prove this algebraically.

$$x_{max} \geq x_1, x_{max} \geq x_2, \cdots, x_{max} \geq x_{10}$$
$$10x_{max} \geq x_1 + x_2 + \cdots + x_{10}$$
$$x_{max} \geq \frac{x_1 + x_2 + \cdots + x_{10}}{10}$$
$$x_{max} \geq \bar{x}$$

Q7 Yes. When ten data elements are ordered, the five-number summary is defined by only six of the values. The minimum is x_1, the maximum is x_{10}, the first quartile is x_3, the third quartile is x_7, and the median is the mean of x_5 and x_6. Moving any of the remaining four values will change the mean with no effect on the box, provided they do not pass any of these above-mentioned points.

Q8 No. The mean formula includes every data point, so changing a single value always affects the mean.

Q9 Yes. If you move two data points the same distance in opposite directions, the sum stays the same, and so does the mean.

Q10 Yes. The median is the mean of x_5 and x_6, so it is only necessary to move or replace one of these two points.

New York City Title I High School Activities with The Geometer's Sketchpad
© 2012 Key Curriculum Press

1. Open **Box and Whiskers Present.gsp.** The sketch has a set of ten data values controlled by ten points on parallel lines. A box-and-whisker summary appears above the points. The actual ordered data are in a column on the left. However, the scale is arbitrary, so you probably will have no use for the actual data.

2. Drag some of the red points to show the class how this affects the image.

Q1 How can you make the box narrower? [Group the data more closely.]

Q2 How can you have a narrow box but long whiskers? [Group most of the data closely, but have one outlier an each side.]

For Q3–Q8, press *Hide Box and Whiskers* before asking. Have students guide your manipulation of the data. Press *Show Box and Whiskers* after the class has reached something approaching a consensus. For each configuration, try to imagine how it could occur with real data. Some suggestions are in the activity notes.

This presentation can be particularly engaging if you give the controls to a student while you and the class direct him or her.

Q3 How can one whisker have zero length? [Group one end of the data on the same value.]

Q4 How can the median fall outside of the box? [This is not possible.]

Q5 Can the box have zero width? [Group the middle part of the data on the same value.]

Q6 Can both whiskers have zero length? [Put the three lowest on the same value and the three highest on the same value.]

3. Press the *Show Mean* button.

Q7 Can the mean fall outside of the box? [Try closely grouped data, but with one extreme outlier.]

Q8 Can the mean be greater than the maximum? [This is not possible.]

For the remaining questions, do not hide the box and whiskers. You will have to be able to see it as the data points are moving.

Q9 Is it possible to move the mean without changing the box, the whiskers, or the median? [Moving any data point will change the mean, but the box and whiskers are unchanged when you move the 2nd, 4th, 7th, or 9th data points.]

Q10 Is it possible to move one point without moving the mean? [It is not.]

Q11 Is it possible to move two points without moving the mean? [Yes, provided they are moved the same distance in opposite directions.]

Box and Whiskers

The *box-and-whiskers plot* (sometimes just called a *box plot*) is a recent development in statistical analysis. You cannot derive any detailed information from it, but it gives you a convenient, easily understood graphical representation of the data distribution.

SKETCH AND INVESTIGATE

1. Open **Box and Whiskers.gsp.**

The sketch contains ten data values represented as points on parallel lines. Above the points are a box and whiskers. You can change a value by sliding its corresponding point right or left. The data are ordered and displayed on the left, but the actual numerical values are not important for this activity.

2. Before answering any of the questions, take a minute to experiment with the sketch. Drag the data points and observe the effect.

Each of the following questions suggests a special shape for the box and whiskers. In each case, state whether it is possible. If it is not possible, explain why not. If it is possible, make a rough sketch of the data points that will create that configuration, and suggest a real data set that might make this happen. The first one is done as an example.

Q1 Can one whisker have zero length?

A1 This will occur if the lower one-fourth of the data points all have the same value. This might happen if there is a lower limit to the data range. Test a group of people to see how far they can throw a heavy weight. Those who cannot even lift the weight will all score zero.

Q2 Can the median fall outside of the box?

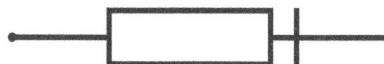

Q3 Can the box have zero width?

Q4 Can both whiskers have zero length?

3. Press the *Show Mean* button. The mean is represented by a green bar. Normally, the mean is not shown on a box-and-whisker plot. It appears here so that you can observe its relationship to the data distribution.

Q5 Can the mean fall outside of the box?

Q6 Can the mean be greater than the maximum?

Questions Q7–Q9 involve moving data points. In doing so, you are actually moving from one data set to another. In real life this could happen when a restaurant manager changes the prices of certain menu items, or when a sports team makes a player trade.

Q7 By moving one or more data points, can you move the mean without changing the box, the whiskers, or the median?

Q8 Can you move a single data point without changing the mean?

Q9 Can you move two data points without changing the mean?

Q10 Can you change the median by moving a single data point?

Lines of Fit

PROCESSING DATA

Q1 Answers will vary. The important thing is that students support their answers with arguments. This is a good question for discussion.

2. To show point labels as they plot the points, students can choose **Edit | Preferences | Text** and set Sketchpad to show labels automatically for all new points.

BEST-FIT LINE

The regression line constructed in this activity is not a least squares regression. Students only have to understand that their objective is to minimize the difference between estimated height and actual measured height.

Q2 In order to fit a line through all of the points, they would have to be collinear, and clearly, they are not. (In fact, it would not be possible to fit any function graph to the points, because two of them have the same *x* value.) However, it *is* possible to draw a line that comes close to all the points.

Q3 Answers will vary. Example: $y = 2.9x + 103$

Q4 The length of a segment represents the amount by which a data point misses the line vertically. It is blue if the point is above the line and red if it is below. The measurements are the lengths of the line segments in coordinate units.

Q5 This should be about the same as the answer to Q3, but this second method is more reliable.

Q6 About 159 cm.

EXPLORE MORE

With students at different stages of physical development, this experiment can generate an interesting range of data. You could also use this data to take a closer look at the question in Q1. Create separate lines of fit for girls and boys, and see if there is a significant difference. (Keep in mind that some students may be sensitive about having their body measurements taken. Modify the directions as needed.)

This problem involves an archaeologist who has uncovered a man's shoe, which would fit a foot of length 19 cm. She would like to get a rough estimate of the man's height. There are eight men and five women working at the dig site. She measures the foot length and the height of each man, but she does not measure the women.

1. Open **Lines of Fit Present.gsp.** Page 1 has a set of coordinate axes at an appropriate scale.

2. Select the table and choose **Graph | Plot Table Data.** This shows a scatter plot representing the foot lengths and heights of eight men.

3. Choose the **Line** tool to draw a line through the data plot. You must click in two places to define the line. Be careful not to click on any existing objects. You want both points to be independent points.

Normally, for such a rough estimate, you would simply draw the line so that it looks right. Adjust the line by dragging the two line points. When the class concurs on its position, find the equation and get an estimate.

4. Select the line. Choose **Measure | Equation.** Have students substitute 19 for x in the equation and solve for y.

5. How good was the line? Go to page 2. This has the same scatter plot. From each point, a vertical line segment represents the residual.

Q1 Ask what the line segments represent. For any foot length (x), the line estimates a man's height (y). Each line segment represents the difference between the actual height of a man and the estimate.

Q2 On the left side of the screen is a measurement labeled *total error*. This is the sum of the absolute values of the residuals. Tell students that this number represents the sum of the lengths of all of the colored line segments. Ask them how to use that number to improve the best-fit line. Help them understand that this number should be as small as possible.

6. Keeping an eye on the sum calculation, alternate between adjusting the two control points of the line to minimize the sum. Use the equation of the new line to make a new estimate of the man's height.

Encourage a discussion of the methods used. Was it right to exclude women from the sample? Should foot length and height have a linear relationship? How could the archaeologist improve the estimate?

It would not take long to conduct the similar measurements on the students.

Lines of Fit

In science you often have to gather data from observations and use them to make a reasonable guess about something you cannot see. If you have a lot of data and they show a consistent pattern, then you can have more confidence in your estimate.

PROCESSING DATA

Suppose you are an archaeologist, and your team has uncovered the remains of a shoe. The foot that fit the shoe would be 19 cm long, and by the type of footwear, you feel certain that it was a man's. In order to form a more complete picture of the shoe's owner, you would like to estimate his height. There are eight men working at the dig site. You decide to use them as your sample. You measure the foot length and height of each man. Here are your data.

Foot length (cm)	25	31	16	21	21	27	28	24
Height (cm)	189	195	149	174	158	169	180	172

Q1 There are also five women working at the site, but you do not measure them. Should you have included them in the sample? Explain.

1. Open **Lines of Fit.gsp.** Notice that the horizontal and vertical scales of the coordinate axes are different.

To label all the points, choose **Edit | Select All** and then choose **Display | Label Points.**

2. Choose **Graph | Plot Points.** Enter 25 in the first box and 189 in the second, then click **Plot.** Repeat this for each of the ordered pairs. When you have plotted the last point, click **Done.** Label all the points.

FINDING A LINE OF FIT

Q2 The simplest way (but not always the best way) to model the relationship between two variables is with a line. Is it possible to draw a line through the eight points on your screen?

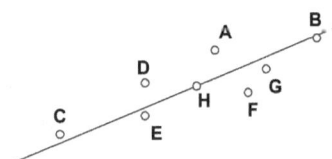

3. Choose the **Line** tool. Click in two places on the screen. When you do this, be careful not to click any of the existing points or axes.

4. Choose the **Arrow** tool. By dragging the two points that define the line, move it into a position where you think it best fits the data points.

This line represents your best guess. If you were to plot the foot length and height of another man, you would expect it to fall somewhere near the line.

5. Select the line. Choose **Measure | Equation.**

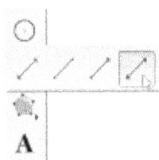

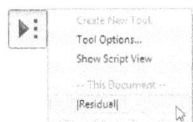

Q3 What is the equation of your line?

6. Press and hold the **Custom** tool icon, and choose |**Residual**| from the menu that appears. Click one of the data points and then click the line. The tool plots a line segment and a measurement. Repeat this at each data point.

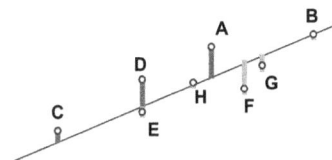

Q4 What is the meaning of the length and color of the line segments that are now attached to the data points? How are the measurements related to the line segments?

Perhaps you can adjust your line to fit the data better. The measurements are the magnitudes of the residuals. They tell you by how much each point misses the line vertically. In order to fit the data well, you need to minimize these values. The trouble is, if you move the line to make one smaller, you are probably making another one bigger.

7. Choose **Number | Calculate.** Compute the sum of the magnitudes of the residuals. This is the total amount by which the line misses the data.

Q5 Adjust the line to minimize that last calculation. What is the new equation for the line? Compare this with your answer to Q3.

8. Construct a point on the line. Measure its coordinates.

Q6 Drag the new point along the line until the *x*-coordinate is about 19. Based on your analysis, what was the approximate height of the man who owned the shoe?

EXPLORE MORE

Using classmates as a data sample, do some measurements and find a line of fit. There are many measurements that you could use—for example, hand span, arm span, and shoulder width.

Wait for a Date

MODEL IN ONE DIMENSION

Q1 If you and your friend wait more than 10 minutes for each other, it is more likely you will meet.

MODEL IN TWO DIMENSIONS

Q2 Points on the diagonal segment represent instances in which you and your friend arrive at the exact same moment. Any placement of point *P* on or very close to this segment guarantees you'll meet. Locations 2 and 3 both satisfy this condition.

Q3 The green points cluster entirely along a diagonal strip, whereas the red points cluster entirely in two congruent triangular regions on either side of the strip. In this illustration, *G* = green and *R* = red.

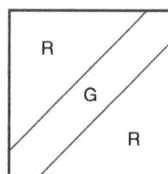

Q4 The diagonal segment connecting (12:00, 12:00) and (1:00, 1:00) represents the set of points for which you and your friend arrive at the exact same time. This is the same diagonal segment that appears in Q2.

Q5 Pick any point on the diagonal segment from Q4, for instance (12:30, 12:30). If your friend arrives 10 minutes after you do, this point is translated vertically by 10 minutes to (12:30, 12:40).

The same pattern holds for every point on the diagonal segment. The collection of all such points is a segment that is a vertical translation by 10 of the diagonal segment. The exact coordinates of the line run from (12:00, 12:10) to (12:50, 1:00). This new segment represents one border of the green region shown in Q3.

Q6 The same reasoning from Q5 applies here. Translating the diagonal segment from Q4 by −10 minutes vertically represents those times when your friend arrives 10 minutes before you do. The exact coordinates of this line run from (12:10, 12:00) to (1:00, 12:50). This segment is a border of the green region shown in Q3.

Q7 If the length of the horizontal and vertical axes is 60 (representing 60 minutes), then the base and height of each red right triangle is 50.

The area of both triangles combined is $50 \times 50 = 2500$. This is 69% of the square, leaving 31% for the green points.

Q8 The portion of the square filled with green points represents the probability that you and your friend will meet. Thus there is a 31% chance of your meeting.

EXPLORE MORE

Q9 Answers will vary.

Q10 The probability that you and your friend will meet is $1 - \left(1 - \frac{t}{60}\right)^2$.

Q11 The lower triangle has base and height of 55 and an area of 1512.5. The upper triangle has base and height of 45 and an area of 1012.5. The red area (the sum of the triangle areas) is 2525, which is 70% of the square. Therefore the probability that you and your friend will meet is 30%. In general, if one person is willing to wait t_A minutes and the other is willing to wait t_B minutes, the probability is

$$1 - \frac{1}{2}\left(1 - \frac{t_A}{60}\right)^2 - \frac{1}{2}\left(1 - \frac{t_B}{60}\right)^2$$

Wait for a Date

You and a friend arrange for a lunch date next week between 12:00 and 1:00 in the afternoon. A week later, however, neither of you remembers the exact meeting time. As a result, each of you arrives at a random time between 12:00 and 1:00 and waits exactly 10 minutes for the other person. When the 10 minutes have passed, each of you leaves if the other person has not come.

What is the probability that the two of you will meet?

MODEL IN ONE DIMENSION

1. Open **Wait for a Date.gsp.** The times when you and your friend arrive are represented as two points, *A* and *B*, at random locations along a timeline. In their initial locations, *A* and *B* arrive 6 minutes apart, well within your 10-minute limit.

2. Press the *Do A and B meet?* button several times. Doing so moves points *A* and *B* to new, random locations along the segment. Notice how the appearance of the stick figures changes depending on the time interval separating them.

3. Drag point *T* to change the amount of time you and your friend will wait. Now run the simulation again several times.

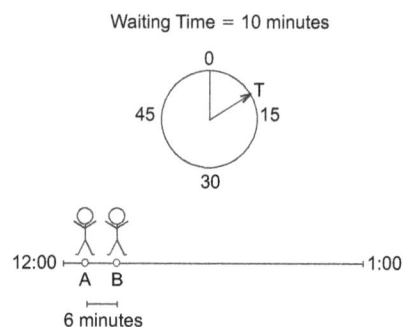

Q1 How can you change the waiting time to make it more likely that you and your friend will meet?

MODEL IN TWO DIMENSION

Viewing a one-dimensional Sketchpad simulation helps you to understand the problem, but it does not allow you to compute the exact probability that you and your friend will meet. A shift in perspective from a one-dimensional to a two-dimensional model makes a big difference, as you will soon discover.

4. Open the second page of the sketch. As before, points *A* and *B* are on a horizontal timeline, but now there is a vertical timeline that shows when *B* arrives.

Q2 Point *P* is at the intersection of lines through *A* and *B* that are perpendicular to the two axes. The following picture shows four possible locations of *P*. For which two locations does it appear that you and your friend meet?

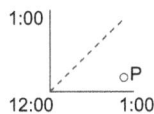 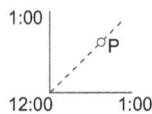 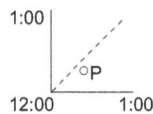 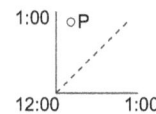

(1)　　　　　　　(2)　　　　　　　(3)　　　　　　　(4)

5. Press the *Do A and B meet?* button. Doing so moves points *A* and *B* to new, random locations along the segment. When *A* and *B* meet, point *P* leaves a green trace on-screen. When the two do not meet, the trace is red.

6. To speed up the process, choose **Display | Show Motion Controller,** and click on the upward-pointing arrow to increase the speed.

To clear the screen, choose **Erase Traces** from the Display menu.

Q3 Run the simulation for a while. Describe the emerging pattern of red and green points.

Q4 Draw the set of all points P for which you and your friend arrive at the exact same time. Describe this set of points.

Q5 Draw the set of all points *P* for which your friend (point *B*) arrives exactly 10 minutes after you (point *A*). Describe this set of points.

Q6 Draw the set of all points *P* for which you arrive exactly 10 minutes after your friend. Describe this set of points.

Use your answers to Q4–Q6 to help with the next two questions.

Q7 What portion of the square formed by the two axes is filled with red points? What portion is filled with green points?

Q8 What is the probability that you and your friend will meet?

EXPLORE MORE

Q9 Drag point T to change the waiting time. Now run the simulation again. What is the probability that you and your friend will meet?

Q10 If you and your friend are willing to wait t minutes for each other, what is the probability you'll meet?

Q11 How do things change if you are willing to wait for 15 minutes, but your friend is willing to wait for only 5 minutes?

5

Lines and Polygons (Geometry)

Introducing Points, Segments, Rays, and Lines

SKETCH AND INVESTIGATE

Q1 You can make the distance between two points zero by dragging one point onto the other.

Q2 The length of a segment is the distance between the segment's endpoints.

Q3 The midpoint traces a straight path. This path is parallel to segment AB and half as long.

Q4 Ray AB could not be called ray BA. Ray BA has endpoint B and passes through point A.

Q5 You can't measure the length of a ray because it is infinitely long.

Q6 You can't find the midpoint of a ray because a ray only has one endpoint and is infinitely long.

Q7 $\overrightarrow{AB}$ and $\overrightarrow{AC}$

Q8 Lines and rays are infinitely long; segments are finite.

A line has no endpoints, a ray has one endpoint, and a segment has two endpoints.

Lines, rays, and segments are all straight.

The notation for a line, ray, or segment uses two letters. Each letter represents a point on that object. In the notation for a ray, one of the points must be an endpoint. In the case of a segment, both of the points must be endpoints. In the case of a line, both points are any (distinct) points on the line.

Q9 $\overrightarrow{AB}$ and $\overrightarrow{BA}$ are rays. $\overline{AB}$ is a segment.

Q10 Students are most likely to use the **Ray** tool to construct $\overrightarrow{AB}$ and $\overrightarrow{AC}$, then adjust point A so that it's between points B and C. This is fine, but it won't remain a line when dragged. If you construct $\overrightarrow{AB}$ and use the same two control points to construct $\overrightarrow{BA}$, the rays will remain a line when dragged.

Introducing Points, Segments, Rays, and Lines

In this activity you'll experiment with drawing, dragging, measuring, and labeling points, segments, rays, and lines. These objects, along with circles, are the building blocks of most geometric constructions.

SKETCH AND INVESTIGATE: POINTS AND SEGMENTS

Note: If at any time you think you've made a mistake or you want to do something differently, you can always undo as many steps as you like. The **Undo** and **Redo** commands are in the Edit menu.

1. Choose the **Point** tool and click in the sketch to construct a point. Click again to construct a second point. Notice that the most recently constructed point is *selected*: It appears with an outline.

2. Choose the **Selection Arrow** tool and click in a blank area in the sketch. This deselects everything.

3. Choose the **Text** tool. Position the finger over a point; then click to display that point's label. Display the other point's label, too.

   By default, point labels start with *A*.

   A   B

4. With the **Selection Arrow** tool, click both points. Now both points should be selected.

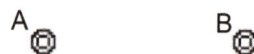

   A   B

5. In the Measure menu, choose **Distance**.

   AB = 2.72 cm

6. Drag one of the points and observe the measurement.

   A   B

   Q1 How can you make the distance between the two points zero?

7. Choose the **Segment** tool and draw a segment connecting the two points. You'll see a triple segment at first, indicating that the segment is selected.

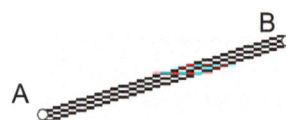

8. With the segment selected, go to the Measure menu and choose **Length**.

9. Use the **Selection Arrow** tool to drag either endpoint of the segment.

   Q2 How does the length of a segment compare to the distance between its endpoints?

10. Use the **Segment** tool to construct a second segment with one endpoint attached to the first segment. To do this, click on a blank area of the sketch, and then click on the original segment.

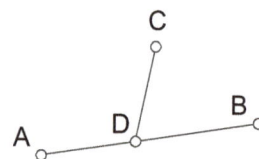

11. Use the **Text** tool to show the labels of this segment's endpoints.

12. Use the **Selection Arrow** tool to drag point D to confirm that it is attached to $\overline{AB}$.

13. Select $\overline{CD}$ (the segment, not its endpoints); then go to the Construct menu and notice what choices are available. Choose **Midpoint**.

14. Click in a blank area to deselect everything.

15. Select point D.

16. In the Edit menu, drag to the Action Buttons submenu and choose **Animation**. You'll get a dialog box you can use to specify animation settings. To choose the default settings, click **OK**. You've created an Animation action button in your sketch.

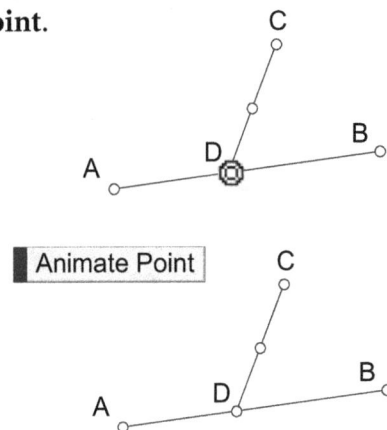

17. Press the action button (by clicking on it) to start the animation.

18. Press the button again to stop the animation.

19. Select the midpoint; then, in the Display menu, choose **Trace Midpoint**.

20. Press the Animation button again and observe the path that the midpoint traces.

Q3 Describe the path that the midpoint traces as point D moves back and forth.

SKETCH AND INVESTIGATE: RAYS AND LINES

21. In the File menu, choose **New Sketch**.

22. Position your pointer on the **Segment** tool, and then click and hold. A palette of **Straightedge** tools will pop out to the right. Drag right and choose the **Ray** tool.

23. Draw a ray in your sketch. Notice that the ray extends in one direction beyond the edge of your sketch window.

24. Use the **Text** tool to show the labels of the ray's control points.

25. Use the **Selection Arrow** tool to drag each point to observe how it controls the ray.

Q4 A ray with endpoint A that passes through a point B is called ray AB (represented symbolically as $\overrightarrow{AB}$). Could it also be called ray BA? Explain.

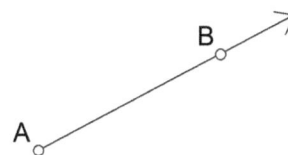

26. Select the ray and go to the Measure menu. Note that **Length** is grayed out.

Q5 Why do you think you can't measure the length of a ray?

27. With the ray still selected, go to the Construct menu and look at your choices. Choose **Point on Ray**.

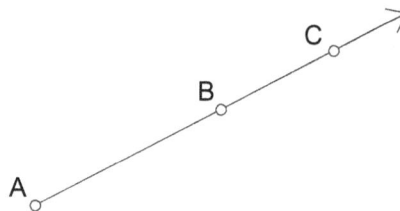

Q6 Why can't you construct the midpoint of a ray?

28. Drag this new point to see how its behavior compares to that of the ray's two control points.

Q7 Give two different names to the ray shown here. Use just two points in each name.

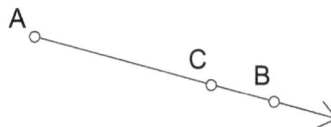

29. Press and hold down the **Ray** tool, and then drag right to choose the **Line** tool.

30. Experiment with drawing lines in your sketch.

Q8 List all the similarities and differences you can between segments, rays, and lines.

Q9 Name two rays and a segment that lie on the line here.

Q10 In Sketchpad, construct a line without using the **Line** tool. Explain what you did. Does your line remain a line when you drag points?

SKETCH AND INVESTIGATE

Q1 The sum of the angles in any triangle is 180°.

Q2 We need to show that the three angles that form a straight angle at point B are congruent to the three angles of the original triangle. Because the horizontal line was constructed parallel to the opposite side of the triangle, the angles marked below with a single arc are congruent because they are alternate interior angles. The same argument holds for the angles marked with two arcs. The third unmarked angle at point B is the remaining angle of the original triangle. So the three angles, which form a straight angle at point B (and whose measures thus add up to 180°), are each congruent to the three angles of the original triangle. Therefore, the three angle measures of the triangle must also add up to 180°.

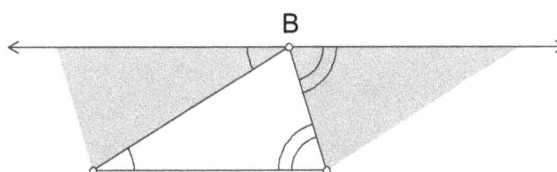

EXPLORE MORE

13. Students should discover that the angle sum for any n-gon is $180°(n - 2)$.

Triangle Sum

This is a two-part investigation. First you'll investigate and make a conjecture about the sum of the measures of the angles in a triangle, and then you'll continue sketching to demonstrate why your conjecture is true.

SKETCH AND INVESTIGATE

To measure an angle, select three points, with the vertex your middle selection. Then, in the Measure menu, choose **Angle**.

Choose **Calculate** from the Number menu to open the Calculator. Click a measurement to enter it into a calculation.

Select point *B* and $\overline{AC}$; then, in the Construct menu, choose **Parallel Line**.

Select the three vertices; then, in the Construct menu, choose **Triangle Interior**.

Double-click the point to mark it as a center. Select the interior; then, in the Transform menu, choose **Rotate**.

Color is in the Display menu.

1. Construct △*ABC*.

2. Measure its three angles.

3. Calculate the sum of the angle measures.

4. Drag a vertex of the triangle and observe the angle sum.

m∠CAB = 43.3°
m∠ABC = 87.2°
m∠BCA = 49.5°

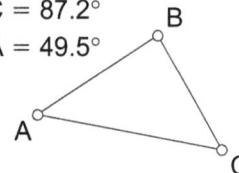

Q1 What is the sum of the angles in any triangle? _____

Follow these steps to investigate why your conjecture is true.

5. Construct a line through point *B* parallel to $\overline{AC}$.

6. Construct the midpoints of $\overline{AB}$ and $\overline{CB}$.

7. Construct the interior of △*ABC*.

8. Mark one of the midpoints as a center for rotation and rotate the interior by 180° about this point.

9. Give the new triangle interior a different color.

10. Mark the other midpoint as a center and rotate the interior by 180° about this point.

11. Give this new triangle interior a different color.

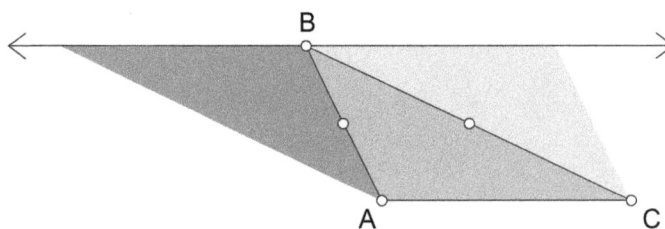

12. Drag point *B* and observe how the three triangles are related to each other and to the parallel line.

Q2 Explain how each of the three angles at point *B* is related to one of the three angles in the triangle. Explain how this demonstrates your conjecture from Q1.

EXPLORE MORE

13. Investigate angle sums in other polygons.

SKETCH AND INVESTIGATE

Q1 The sum of the measures of the two remote interior angles equals the measure of the exterior angle.

Q2 The triangles created by rotation and translation are congruent to the original triangle, so their corresponding angles are also congruent. The two angles that fill the exterior angle correspond to the two remote interior angles of the original triangle, as marked below, so this shows that the measures of the remote interior angles sum to the measure of the exterior angle.

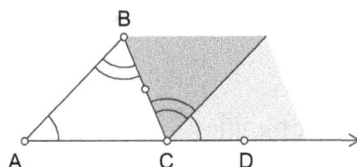

Exterior Angles in a Triangle

An *exterior angle* of a triangle is formed when one of the sides is extended. An exterior angle lies outside the triangle. In this investigation you'll discover a relationship between an exterior angle and the sum of the measures of the two remote interior angles.

SKETCH AND INVESTIGATE

Press and hold the pointer on the **Segment** tool and drag right to choose the **Ray** tool.

1. Construct $\triangle ABC$.

2. Construct $\overrightarrow{AC}$ to extend side $\overline{AC}$.

3. Construct point D on $\overrightarrow{AC}$, outside of the triangle.

$m\angle CAB = 45.5°$
$m\angle ABC = 67.2°$
$m\angle BCD = 112.7°$

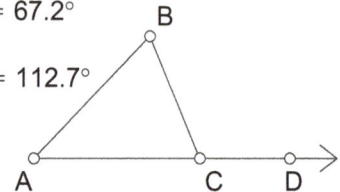

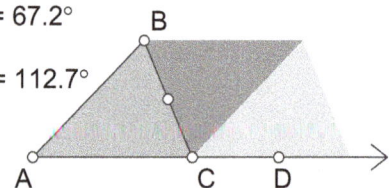

Select, in order, points B, C, and D. Then, in the Measure menu, choose **Angle**.

4. Measure exterior angle BCD.

5. Measure the remote interior angles $\angle ABC$ and $\angle CAB$.

6. Drag parts of the triangle and look for a relationship between the measures of the remote interior angles and the exterior angle.

Choose **Calculate** from the Number menu to open the Calculator. Click a measurement to enter it into a calculation.

Q1 How are the measures of the remote interior angles related to the measure of the exterior angle? Use the Calculator to create an expression that confirms your conjecture.

Follow these steps to see why your conjecture is true.

Select the vertices; then, in the Construct menu, choose **Triangle Interior**.

7. Construct the interior of $\triangle ABC$.

8. Mark AC as a vector.

Select point A and point C in order; then, in the Transform menu, choose **Mark Vector**.

9. Translate the interior by the marked vector.

$m\angle CAB = 45.5°$
$m\angle ABC = 67.2°$
$m\angle BCD = 112.7°$

Select the interior; then, in the Transform menu, choose **Translate**.

10. Give the new triangle interior a different color.

11. Construct the midpoint of $\overline{BC}$.

Double-click the point to mark it as a center. Select the interior; then, in the Transform menu, choose **Rotate**.

12. Mark the midpoint as a center for rotation and rotate the triangle interior about this point by 180°.

13. Give this new triangle interior a different color.

Q2 Explain how the two angles that fill the exterior angle are related to the remote interior angles in the triangle. Explain how this demonstrates your conjecture from Q1.

Exterior Angles in a Polygon

SKETCH AND INVESTIGATE

Q1 The sum of the measures of the exterior angles in any convex polygon is 360°. Make sure students try this investigation with other polygons in addition to the pentagon shown in the example. They could actually manipulate their pentagon into a quadrilateral or triangle, making one or more angle measures disappear.

If students print out their polygons with the exterior angles, they can actually cut out the exterior angles and rearrange them around a single point to show that they sum to 360°.

Q2 As you dilate the figure toward any point, the polygon will shrink toward that point, leaving you with only the angles surrounding a common vertex. This has the same visual effect as zooming out from the polygon.

When the polygon appears to be the size of a point, the exterior angles appear as spokes radiating from the point, with the angle markers forming a circle around the point. This provides a visual proof that their sum is 360°, or the total number of degrees in one revolution around a point.

EXPLORE MORE

11. Interestingly, this same conjecture applies to concave polygons, too, if you consider the "exterior" angles that fall inside the polygon to be negative. To investigate this with Sketchpad, be sure to set Preferences to show angle measures in directed degrees. The figure below demonstrates how selection order determines the sign of an angle measure. If you think in terms of a rotation from one ray to the other, a counterclockwise rotation is positive and a clockwise rotation is negative.

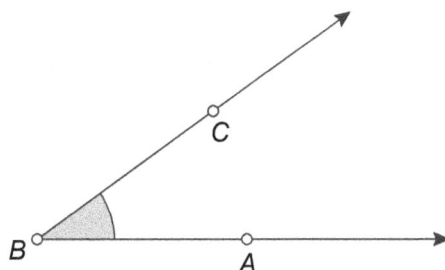

$m\angle ABC = 35.2°$
$m\angle CBA = -35.2°$

Exterior Angles in a Polygon

An exterior angle of a polygon is formed when one of the sides is extended. Exterior angles lie outside a convex polygon. In this investigation you'll discover the sum of the measures of the exterior angles in a convex polygon.

Do this investigation with a triangle, a quadrilateral, or a pentagon. Plan together with classmates at nearby computers to investigate different polygons so that you can compare your results. The activity here shows a pentagon. Don't let that throw you if you're investigating a triangle or a quadrilateral—the basic steps are the same.

SKETCH AND INVESTIGATE

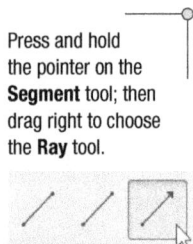

Press and hold the pointer on the **Segment** tool; then drag right to choose the **Ray** tool.

1. Use the **Ray** tool to construct a polygon with each side extended in one direction. Be sure to construct the polygon without creating any extra points. Your initial sketch should have the same number of points (vertices) as sides. If your polygon didn't end up convex, drag a vertex to make it convex.

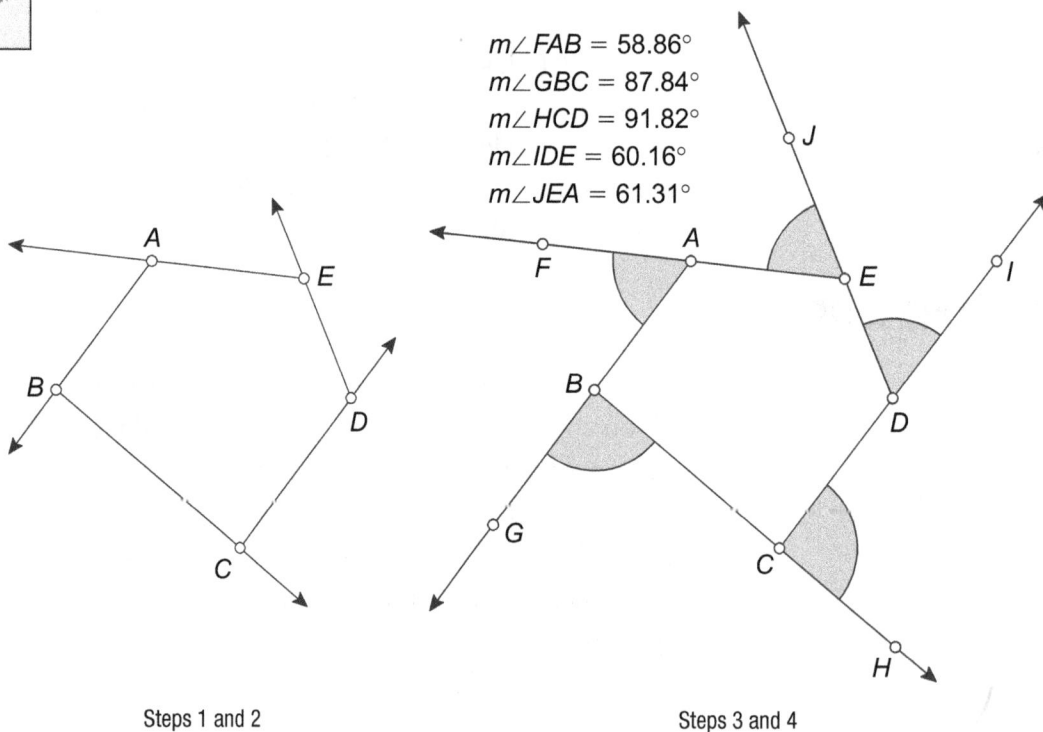

$m\angle FAB = 58.86°$
$m\angle GBC = 87.84°$
$m\angle HCD = 91.82°$
$m\angle IDE = 60.16°$
$m\angle JEA = 61.31°$

Steps 1 and 2   Steps 3 and 4

2. Use the **Text** tool to label the vertices of the polygon.

3. Create an angle marker in each external angle by dragging the **Marker** tool counterclockwise from one side of the angle to the other.

4. Select all the angle markers and choose **Measure | Angles**.

New York City Title I High School Activities with The Geometer's Sketchpad
© 2012 Key Curriculum Press

Choose **Calculate** from the Number menu to open the Calculator. Click a measurement to enter it into a calculation.

5. Calculate the sum of the exterior angles.

6. Drag different vertices of your polygon and observe the angle measures and their sum. Be sure the polygon stays convex.

7. Compare your observations with those of classmates who did this investigation with different polygons.

Q1 Write a conjecture about the sum of the measures of the exterior angles in any polygon.

Follow the steps below for another way to demonstrate this conjecture.

Double-click a point to mark it as a center.

In the Edit menu, choose **Select All.** Then click each measurement to deselect it.

8. Mark any point in the sketch as a center for dilation.

9. Select everything in the sketch except for the measurements.

Press and hold the pointer on the **Arrow** tool, and then drag right to choose the **Dilate Arrow** tool.

10. Change your **Arrow** tool to the **Dilate Arrow** tool and use it to drag any part of the construction toward the marked center. Keep dragging until the polygon is nearly reduced to a single point.

Q2 Write a paragraph explaining how this demonstrates the conjecture you made in Q1.

EXPLORE MORE

In the Edit menu, choose **Preferences** and go to the Units panel. In the Angle Units pop-up menu, choose **directed degrees.**

11. Investigate the sum of the exterior angle measures in concave polygons. For this investigation, you may want to measure angles in directed degrees. The sign of an angle measured in directed degrees depends on whether the angle is marked in a clockwise or counterclockwise direction.

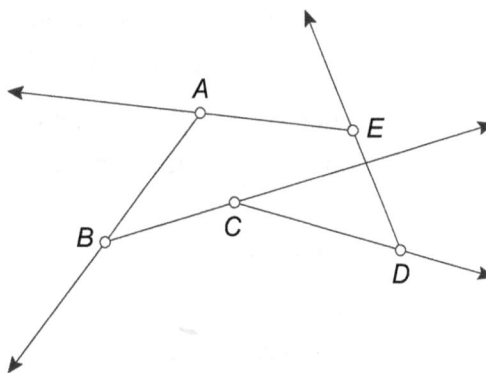

Polygon Angle Measure Sums

SKETCH AND INVESTIGATE

Q1 No. The angle sums should remain constant as long as the polygon is convex.

Q2 The angle sum increases by 180° with each additional side.

Q3 The diagonal of the quadrilateral divides the quadrilateral into two triangles. Each triangle has an angle sum of 180°, so the sum of the angle measures in the quadrilateral is 2(180°) = 360°.

In the case of the pentagon, the two diagonals create three triangles. So the sum of the angle measures in the pentagon is 3(180°) = 540°.

Q4 The sum of the angle measures of an n-gon is $(n - 2)180°$. $n - 2$ is the number of triangles created when you "triangulate" the n-gon.

EXPLORE MORE

Q5 When the polygon is concave, Sketchpad's calculation of the sum is no longer constant; it becomes less than $(n - 2)180°$. When it encounters a concave angle, the angle marker in Sketchpad flips so that it's marking the smaller angle: the angle outside the polygon, which is less than 180°. Students may conjecture that the sum would remain constant if the interior angle (the *reflex angle*) were being measured. Students may also reason that a concave n-gon can still be divided into $n - 2$ triangles. The hexagon shown at right is concave at vertex B but still contains four triangles, showing that its angles total 720°.

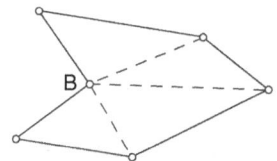

Q6 When the calculation includes the interior (reflex) angle, the sum is constant. The sum remains constant for a polygon with two concave vertices, suggesting that this method for finding the sum works for all simple polygons.

Q7 For a crossed polygon, it's not possible to unambiguously distinguish the interior from the exterior. The sum of the angles depends on the nature of the crossings.

Q8 If you turn the polygon completely inside out, so that the angle markers are on the outside of the polygon rather than the inside, the sum of the angles is now $(n + 2)180°$. (Note that these are not the exterior angles of the polygon.)

Polygon Angle Measure Sums

You may already know what the sum of the angle measures is in any triangle. In this activity you'll see how that sum is related to the sum of the angle measures in other polygons.

SKETCH AND INVESTIGATE

1. Use the **Segment** tool to draw a triangle, a quadrilateral, and a pentagon across the top of your sketch.

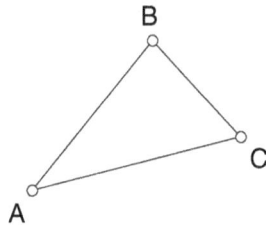

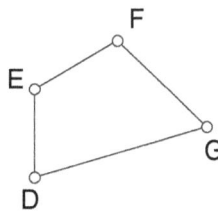

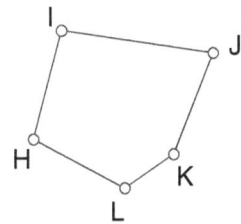

To measure an angle, press the **Marker** tool on the vertex, drag to the interior of the angle, and release near the left side of the angle. (The arrowhead should point counterclockwise.) Then, with the angle marker still selected, choose **Measure | Angle.**

2. Measure each of the three angles in the triangle and arrange the measurements under the triangle.

3. Measure the four angles of the quadrilateral and arrange them under the quadrilateral.

4. Measure the five angles of the pentagon and arrange them under the pentagon.

5. Calculate the sum of the triangle angle measures.

6. Drag any vertex of the triangle and observe the angle measures and the sum.

Choose **Number | Calculate** to open the Calculator. Click a measurement to enter it into a calculation. If the Calculator is in the way of your measurements, move it by dragging the title bar.

7. Calculate the sum of the quadrilateral angle measures.

8. Drag any vertex of the quadrilateral and observe the angle measures and the sum. Be sure to keep the quadrilateral convex.

9. Calculate the sum of the pentagon angle measures.

10. Drag any vertex of the pentagon and observe the angle measures and the sum. Be sure to keep the pentagon convex.

Q1 Did any of the sums change when you dragged (as long as the polygons were convex)?

Q2 Describe the pattern in the angle measure sums as you increased the number of sides in a polygon.

11. Draw a diagonal in the quadrilateral.

12. Draw two diagonals from one vertex in the pentagon.

Q3 Write a paragraph explaining what these diagonals have to do with the pattern you described in Q2.

Q4 Write an expression for the sum of the angle measures in an *n*-gon.

Q5 Drag one of your vertices so that the polygon is concave. What happens to the sum of the polygon's angles? What happens to the angle marker on the concave angle? Make a conjecture about the sum of the angle measures for a concave polygon.

To change the properties of an angle marker, select it and choose **Edit | Properties.** On the Marker panel, change the angle definition to **counterclockwise.**

Q6 To test your conjecture, change the properties of the angle marker on the concave angle so that it shows the counterclockwise angle rather than the simple angle. What do you conclude about your conjecture? Test your conjecture again, this time using a polygon with two concave angles. What is your conclusion?

Q7 So far, you've investigated both convex and concave simple polygons. (A *simple polygon* is one in which the sides don't cross each other.) What happens if the sides do cross? What can you say about the sum of the angle measures of a crossed polygon?

Q8 What happens if you turn the polygon completely inside out, so that the angle markers are on the outside of the polygon rather than the inside?

New York City Title I High School Activities with The Geometer's Sketchpad
© 2012 Key Curriculum Press

SKETCH AND INVESTIGATE

Q1   a.   Lines with positive slope rise from left to right. Lines with negative slope fall from left to right.

b.   Zero

c.   Steeper lines have slopes with greater absolute value.

d.   Undefined

Q2   Check students' work.

EXPLORE MORE

12.   Students' drawings should resemble the figure below. They should calculate slopes of 2/3 and −2/3.

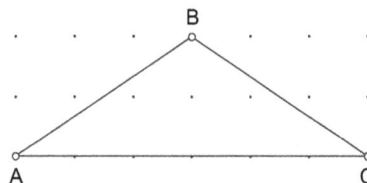

13.   Students should informally explain that slope is calculated by dividing the difference in *y*-coordinates by the difference in corresponding *x*-coordinates. Formally stated, if two points have coordinates (x_1, y_1) and (x_2, y_2), the slope of the line through them is $(y_2 - y_1)/(x_2 - x_1)$.

Different Slopes: The Slope of a Line

If you are a skier, you might describe the slope of a ski hill. If you are a carpenter, you might describe the slope of a roof you built. An economist might describe the slope of a graph. In this activity you'll discover how to recognize the slopes of various lines. You will also play a game, with a partner, that challenges you to recognize various slopes.

SKETCH AND INVESTIGATE

Press and hold the pointer on the **Segment** tool to show the **Straightedge** tool palette. Drag right to choose the **Line** tool.

Select the line; then, in the Measure menu, choose **Slope.**

1. In Preferences, set Distance Units to **cm**.

2. In the Graph menu, choose **Show Grid**.

3. Draw any line.

4. Measure its slope.

5. Drag one of the line's control points and observe the effect on the line's slope.

Slope $\overleftrightarrow{AB}$ = −1.82

6. Drag the line itself and observe the effect on its slope.

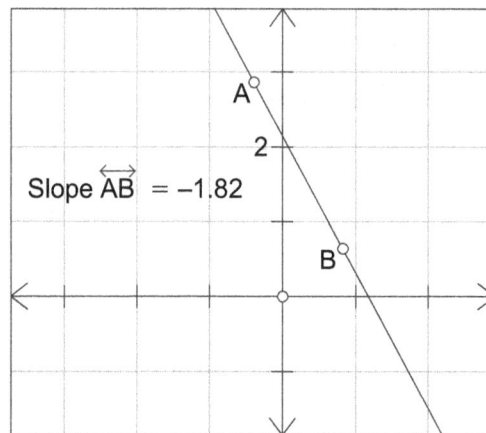

Q1 Continue to change the slope of your line. Investigate the following questions to prepare yourself for the Slope Game:

    a. Which lines have a positive slope and which have a negative slope?

    b. What is the slope of a horizontal line?

    c. How can you tell a steeper slope from a shallower slope?

    d. What is the slope of a vertical line?

PLAYING THE SLOPE GAME

Play this game with a partner.

7. Draw five different random lines in your sketch. Make sure their labels are not showing.

Choose the **Line** tool. In the Edit menu, choose **Select All Lines.** Now you can measure all the slopes at once.

8. Measure the slopes of the five lines.

9. Challenge your partner to match each measured slope with a line. Your partner is allowed to drag only measurements, to move them next to the lines they match. The lines and the points are off limits until all the measurements have been matched up with lines.

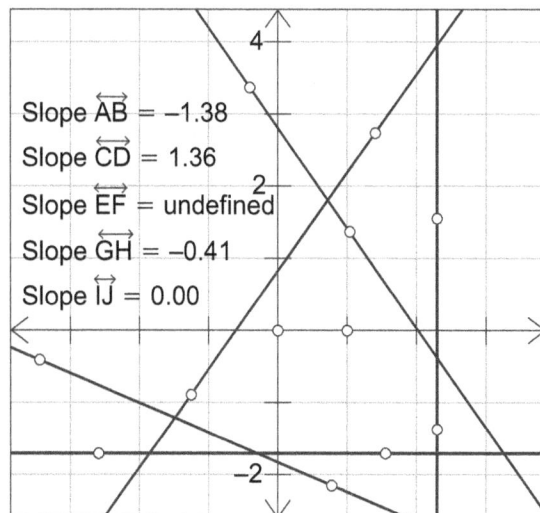

Slope $\overleftrightarrow{AB}$ = −1.38
Slope $\overleftrightarrow{CD}$ = 1.36
Slope $\overleftrightarrow{EF}$ = undefined
Slope $\overleftrightarrow{GH}$ = −0.41
Slope $\overleftrightarrow{IJ}$ = 0.00

Click in a blank area to make sure nothing is selected, and then choose the **Point** tool. In the Edit menu, choose **Select All Points.** Then, in the Display menu, choose **Show Labels.**

10. To check how you scored, show all the point labels. Award one point for each correctly matched slope.

11. Switch roles, scramble the lines, and play the game again. Add a few more lines to make the game more challenging. Measure the slopes of the new lines.

Q2 Record your scores on a separate sheet.

EXPLORE MORE

12. The *rise* of a sloped roof is the vertical distance between its lowest point and its highest point. The *run* is the horizontal distance between the lowest and highest points. Carpenters describe the slope of a roof by comparing the rise and the run. Sketch a scale drawing of a two-sided slanted roof that has a rise of 2 and a run of 3. Calculate the slopes of the two sides of the roof.

13. You will not always need to rely on a computer to measure slopes. Experiment with the sketch **Different Slopes.gsp.** Calculate the slope of the line using the *x*- and *y*-coordinates shown in the sketch. Drag points in the sketch to see how they affect your calculation. Check your calculation by measuring the slope of the line. Explain how you calculated the slope.

SKETCH AND INVESTIGATE

Q1 *EF* equals the average of *AB* and *CD*.

Q2 The midsegment of a trapezoid is parallel to the bases. Its length is the average of the lengths of the bases.

Q3 The midsegment of a triangle is parallel to the third side and is half its length.

EXPLORE MORE

15. Students should discover $A = mh$, where A is the area, m is the length of the midsegment, and h is the height perpendicular to the midsegment. This formula is nice and general and works for both a triangle and a trapezoid.

16. Students should find that the length of $\overline{A'C'}$ is one-third the length of $\overline{AC}$. The segments are parallel to each other. In general, if you dilate points A and C by a scale factor of k, then $A'C' = k(AC)$, and the segments are parallel.

17. The four smaller triangles are similar to the original triangle and congruent to one another. Each has side lengths one-half the original triangle's side lengths and each has area one-fourth the original.

18. Successive triangles get smaller and smaller, but each smaller triangle is still similar to the original.

Midsegments of a Trapezoid and a Triangle

A *midsegment* in a trapezoid connects the midpoints of the two nonparallel sides. In a triangle, the midsegment connects the midpoints of any two sides. In this investigation you'll construct a trapezoid and its mid-segment and discover some properties of the midsegments of a trapezoid. Then you'll apply these properties to the special case of a triangle.

SKETCH AND INVESTIGATE

1. Construct $\overline{AB}$.

2. Construct point *C* not on $\overline{AB}$.

Select point *C* and $\overline{AB}$; then, in the Construct menu, choose **Parallel Line.**

3. Construct a line through point *C* parallel to $\overline{AB}$.

4. Construct $\overline{CD}$, where point *D* is a point on the parallel line.

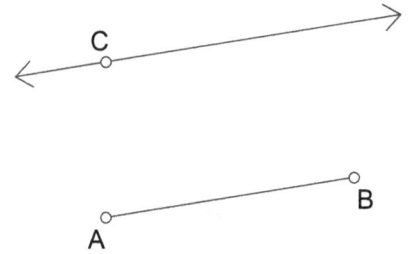

Steps 1–3

5. Construct $\overline{AC}$ and $\overline{DB}$, the legs of your trapezoid.

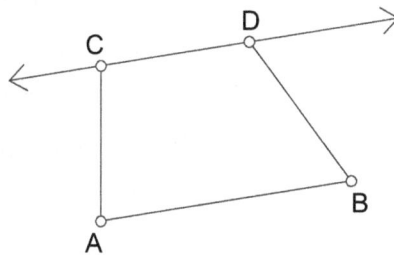

Steps 4 and 5

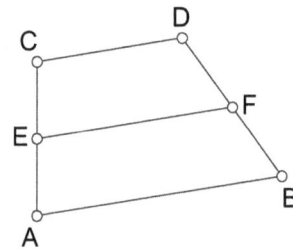

Steps 6–8

Select the line; then, in the Display menu, choose **Hide.**

6. Hide the parallel line.

To measure all three lengths at once, select all three segments. Then, in the Measure menu, choose **Length.**

7. Construct points *E* and *F*, the midpoints of $\overline{AC}$ and $\overline{DB}$.

8. Construct $\overline{EF}$, the midsegment of trapezoid *ACDB*.

9. Measure *AB*, *EF*, and *CD*.

Notice that *EF* is some number between *AB* and *CD*.

10. Drag various parts of the trapezoid and look for a relationship among the lengths of the midsegment and the bases.

Choose **Calculate** from the Number menu to open the Calculator. Click a measurement to enter it into a calculation.

11. Use the measurements for *AB* and *CD* to calculate an expression equal to *EF*.

Q1 Write a conjecture about the midsegment of a trapezoid.

12. Measure the slopes of $\overline{AB}$, $\overline{EF}$, and $\overline{CD}$.

Q2 Write a conjecture about the slope of the midsegment of a trapezoid.

13. Drag point *D* until *CD* is as close to 0 as you can make it. Now you have a triangle.

14. Drag points *A* and *B*. Observe the relationship between *AB* and *EF* and observe the relationship between the slopes.

Q3 Make a conjecture about a midsegment of a triangle.

EXPLORE MORE

15. Come up with an area formula for a trapezoid and a triangle that uses the length of a midsegment.

16. In a new sketch, draw △*ABC*. Mark point *B* as a center and dilate points *A* and *C* by a scale factor of 1/3 (or some other scale factor). Construct $\overline{A'C'}$. How do the length and direction of $\overline{A'C'}$ compare with those of $\overline{AC}$?

17. In a new sketch, construct a triangle with all three midsegments. This divides the triangle into four smaller triangles. Investigate the properties of these triangles.

18. Create a custom tool for constructing a triangle and the midpoints of its sides. Use this tool on the midpoints of the original triangle, and then on the midpoints of the newly constructed triangle, and so on. Make conjectures about the smaller successive midpoint triangles. (You can also do this exploration using **Iterate** in the Transform menu.)

Midpoint Quadrilaterals

SKETCH AND INVESTIGATE

Q1 The quadrilateral whose sides connect the midpoints of any quadrilateral is a parallelogram. The measurements support this conjecture because they show that the opposite sides of the midpoint quadrilateral are equal in length and that opposite sides have equal slope (and therefore are parallel).

Q2 A diagonal divides the quadrilateral into two triangles. Two sides of the midpoint quadrilateral are midsegments of these triangles. This means they are both parallel to the diagonal and half as long. If one pair of opposite sides of a quadrilateral are both equal in length and parallel, the quadrilateral is a parallelogram. (Students might construct the other diagonal and use a second pair of triangles to show that the other pair of sides of the midpoint quadrilateral are also equal in length and parallel.)

EXPLORE MORE

9. A midpoint quadrilateral of a midpoint quadrilateral is still just a parallelogram. Successive midpoint quadrilaterals are alternately similar; that is, the third midpoint quadrilateral is a parallelogram similar to the first, the fourth is similar to the second, and so on. These parallelograms converge on the point of intersection of segments connecting midpoints of opposite sides.

10. The area of the midpoint quadrilateral is half the area of the original quadrilateral. (As an extra challenge, ask students to prove this is true.)

11. The conditions under which a midpoint quadrilateral is a special parallelogram are not obvious. In general, the midpoint quadrilateral of a trapezoid is a parallelogram, that of an isosceles trapezoid is a rectangle, that of a parallelogram is a parallelogram, that of a kite is a rectangle, that of a rhombus is a rectangle, that of a rectangle is a rhombus, and that of a square is a square. See the activity Special Midpoint Quadrilaterals for more discussion.

12. The page "Special Midpoint Quads" of **Midpoint Quadrilaterals Work.gsp** illustrates the most general quadrilaterals whose midpoint quadrilaterals are special parallelograms. This question is the subject of the activity Special Midpoint Quadrilaterals. The midpoint quadrilateral of any quadrilateral whose diagonals are equal is a rhombus. The midpoint quadrilateral of any quadrilateral whose diagonals are perpendicular is a rectangle. The midpoint quadrilateral of any quadrilateral whose diagonals are equal and perpendicular is a square.

Midpoint Quadrilaterals

In this investigation you'll discover something surprising about the quadrilateral formed by connecting the midpoints of another quadrilateral.

SKETCH AND INVESTIGATE

If you select all four sides, you can construct all four midpoints at once.

1. Construct quadrilateral *ABCD*.

2. Construct the midpoints of the sides.

3. Connect the midpoints to construct another quadrilateral, *EFGH*.

4. Drag vertices of your original quadrilateral and observe the midpoint quadrilateral.

5. Measure the four side lengths of this midpoint quadrilateral.

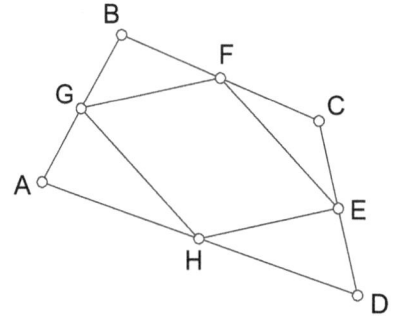

Q1 Measure the slopes of the four sides of the midpoint quadrilateral. What kind of quadrilateral does the midpoint quadrilateral appear to be? How do the measurements support that conjecture?

6. Construct a diagonal.

7. Measure the length and slope of the diagonal.

8. Drag vertices of the original quadrilateral and observe how the length and slope of the diagonal are related to the lengths and slopes of the sides of the midpoint quadrilateral.

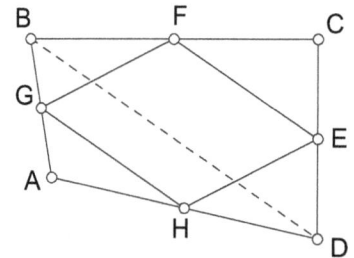

Q2 The diagonal divides the original quadrilateral into two triangles. Each triangle has as a midsegment one of the sides of the midpoint quadrilateral. Use this fact and what you know about the slope and length of the diagonal to write a paragraph explaining why the conjecture you made in Q1 is true.

EXPLORE MORE

9. Construct the midpoint quadrilateral of the midpoint quadrilateral. Then construct *its* midpoint quadrilateral. Do this two or three more times. Describe any patterns you see in the midpoint quadrilaterals.

10. Construct the polygon interiors of a quadrilateral and its midpoint quadrilateral. Measure their areas. Make a conjecture about these areas.

11. What's the midpoint quadrilateral of a trapezoid? An isosceles trapezoid? A parallelogram? A kite? A rhombus? A rectangle? A square? Organize and explain your findings.

12. Under what conditions is a midpoint quadrilateral a rectangle? A rhombus? A square? See if you can construct the most general quadrilateral whose midpoint quadrilateral is one of these.

6

Constructions and Loci (Geometry)

For **GSP5** ACTIVITY NOTES

The more time you give students, the more methods they will come up with. Have students drag vertices of their figures to make sure their constructions are correct. Isosceles triangles that fall apart and can turn into other shapes are underconstrained. A construction that stays an isosceles triangle but that can't take on all possible shapes of an isosceles triangle is overconstrained. Here are various methods for constructing an isosceles triangle that is neither overconstrained nor underconstrained.

Method: Construct a circle AB and radii AB and AC. Construct $\overline{BC}$.
Property: An isosceles triangle has two equal sides.

Method: Construct a line AB and a point C not on the line. Reflect point C across the line. Triangle ACC' is isosceles.
Property: An isosceles triangle has reflection symmetry.

Method: Construct $\overline{AB}$ and its midpoint C. Construct a perpendicular through point C. Construct point D on the perpendicular. Triangle ABD is isosceles.
Property: The perpendicular bisector of the base of an isosceles triangle passes through the vertex of the vertex angle.

Method: Construct $\overrightarrow{AB}$ and $\overrightarrow{AC}$. Construct the angle bisector of $\angle BAC$. Construct point D on the angle bisector and a line through point D perpendicular to the angle bisector. Construct the points of intersection E and F of this line with $\overrightarrow{AB}$ and $\overrightarrow{AC}$. Triangle AEF is isosceles.
Property: The angle bisector of the vertex angle of an isosceles triangle is perpendicular to the base.

Method: Construct $\overrightarrow{AB}$, $\overrightarrow{BA}$, and $\overrightarrow{AC}$. Select points C, A, and B and, in the Transform menu, choose **Mark Angle.** Mark point B as a center. Rotate $\overrightarrow{BA}$ by the marked angle. Construct the point of intersection, D, of this ray and $\overrightarrow{AC}$. Triangle ABD is isosceles.
Property: The base angles of an isosceles triangle are equal in measure.

Constructing Isosceles Triangles

How many ways can you come up with to construct an isosceles triangle? Try methods that use just the freehand tools (those in the Toolbox) and also methods that use the Construct and Transform menus. Write a brief description of each construction method along with the properties of isosceles triangles that make that method work.

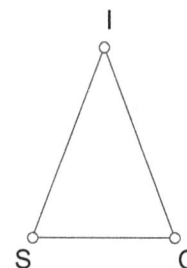

Method 1:

Properties:

Method 2:

Properties:

Method 3:

Properties:

Method 4:

Properties:

New York City Title I High School Activities with The Geometer's Sketchpad

The more time you give students, the more methods they'll come up with. Have students drag vertices of their figures to make sure their constructions are correct. Parallelograms that fall apart and can turn into other shapes are underconstrained. Constructions that remain parallelograms but that can't take on all the shapes of a parallelogram are overconstrained. Here are various methods for constructing a parallelogram that is neither overconstrained nor underconstrained.

Method: Construct segments AB and AC. Construct a line through point B parallel to $\overline{AC}$ and a line through point C parallel to $\overline{AB}$. Construct the intersection of these lines, then hide the lines. Construct the missing segments. (This method is described in more detail in the student pages of the activity Properties of a Parallelogram.)
Properties: This uses the definition of a parallelogram, which states that opposite sides are parallel.

Method: Construct a triangle. Construct the midpoint of one of its sides. Mark the midpoint as a center of rotation and rotate the triangle about this point by 180°.
Properties: A parallelogram has rotational symmetry about the midpoints of its diagonals. Also, this uses the property that a parallelogram consists of two congruent triangles, separated by a diagonal.

Method: Construct a segment and a point not on the segment. Mark the endpoints of the segment as a vector. Translate the point not on the segment by the marked vector. Construct the missing sides.
Properties: A parallelogram has one pair of opposite sides that are both parallel and congruent.

Method: Construct a segment and a point not on the segment. Select the segment and the point and, in the Construct menu, choose **Circle By Center+Radius**. Construct a line through the point, parallel to the segment. Find the point where the circle and the line intersect. This is the fourth vertex of the parallelogram.
Properties: A parallelogram has one pair of opposite sides that are both parallel and congruent.

Method: Construct a segment and its midpoint. Construct a line through the segment attached at the midpoint. Construct a circle centered at the midpoint. Find the two points where the circle intersects the line. These two points and the original endpoints of the segment are the vertices of the parallelogram.

Properties: The diagonals of a parallelogram bisect each other.

Method: Construct a pair of concentric circles. Construct two lines that both contain the center point. Find the two points of intersection of one of the lines with one of the circles, and of the other line with the other circle. These four points are the points of intersection of the parallelogram.

Properties: The diagonals of a parallelogram bisect each other.

Constructing Parallelograms

How many ways can you come up with to construct a parallelogram? Try methods that use the Construct menu, the Transform menu, or combinations of both. Consider how you might use diagonals. Write a brief description of each construction method along with the properties of parallelograms that make that method work.

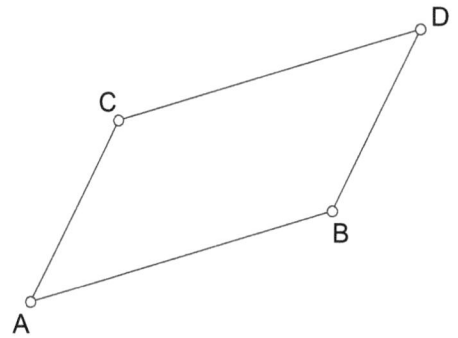

Method 1:

Properties:

_____

Method 2:

Properties:

_____

Method 3:

Properties:

_____

Method 4:

Properties:

Constructing Rectangles

The more time you give students, the more methods they'll come up with. Have students drag vertices of their figures to make sure their constructions are correct. Rectangles that fall apart and can turn into other shapes are underconstrained. Constructions that remain rectangles but that can't take on all the shapes of a rectangle are overconstrained. Here are various methods for constructing a rectangle that is neither overconstrained nor underconstrained.

Method: Construct a segment. Construct perpendicular lines through the endpoints. From a point on one of the perpendiculars, construct a perpendicular line (or a line parallel to the original segment).
Properties: A rectangle is a quadrilateral with four right angles.

Method: Construct a segment AB. Construct a line through point A, perpendicular to $\overline{AB}$. Construct $\overline{AC}$ on this line. Construct a circle with center C and radius AB. Construct a circle with center B and radius AC. Construct $\overline{BD}$ and $\overline{CD}$, where D is the intersection of the circles.
Properties: A rectangle has at least one right angle and the opposite sides are congruent.

Method: Construct a segment AB and its midpoint, C. Construct circle CA. Construct ray $\overrightarrow{DC}$, where point D is on the circle. Construct point E, the intersection of the ray and the circle. Construct $\overline{AD}$, $\overline{DB}$, $\overline{DE}$, and $\overline{EA}$.
Properties: The diagonals of a rectangle are equal and bisect each other.

Method: Construct a segment AB and its midpoint C. Construct a line through point C, perpendicular to $\overline{AB}$. Construct circle CD, where point D is on the perpendicular line. Construct point E, the other intersection of the circle with the line. Construct lines through points D and E, perpendicular to $\overline{DE}$ (or parallel to $\overline{AB}$). Construct lines through points A and B, perpendicular to $\overline{AB}$ (or parallel to $\overline{DE}$).
Properties: The segments connecting midpoints of opposite sides of a rectangle bisect each other and are perpendicular to the sides of the rectangle. A rectangle has two perpendicular lines of symmetry.

Method: Construct a segment AB. Construct a line perpendicular to $\overline{AB}$, through point A. Construct $\overline{AC}$ on this line. Construct $\overline{CB}$ and point D, the midpoint of $\overline{CB}$. Rotate A, $\overline{AB}$, and $\overline{AC}$ 180° about point D.
Properties: A rectangle has at least one right angle and has 180° rotation symmetry.

Method: Construct two perpendicular lines. Construct a point not on either line. Reflect it across one of the lines. Now reflect the two points across the other line.
Properties: A rectangle has two perpendicular lines of symmetry.

Different methods are interesting because they yield different control points for manipulating the rectangle. The most efficient custom tool can be made for the last method, although that tool has three givens. Each of these methods has a corresponding example custom tool.

Constructing Rectangles

How many ways can you come up with to construct a rectangle? Try methods that use the Construct menu, the Transform menu, or combinations of both. Consider how you might use diagonals. Write a brief description of each construction method along with the properties of rectangles that make that method work.

C _____ D

A _____ B

Method 1:

Properties:

_____

Method 2:

Properties:

_____

Method 3:

Properties:

_____

Method 4:

Properties:

New York City Title I High School Activities with The Geometer's Sketchpad
© 2012 Key Curriculum Press

Constructing Rhombuses

The more time you give students, the more methods they'll come up with. Have students drag vertices of their figures to make sure their constructions are correct. Rhombuses that fall apart and can turn into other shapes are underconstrained. Constructions that remain rhombuses but that can't take on all the shapes of a rhombus are overconstrained. Here are various methods for constructing a rhombus. The first is a popular one that is overconstrained. The rest are neither overconstrained nor underconstrained.

Method: Construct circles AB and BA. Construct a rhombus connecting the centers and the two points of intersection of the circles. (This is a special rhombus, composed of two equilateral triangles.)
Properties: A rhombus has four equal sides.

Method: Construct circles AB and BA. Construct circle CA, where point C is a point on circle AB. Construct point D at the point of intersection of circles BA and CA. $ABDC$ is a rhombus.
Properties: A rhombus has four equal sides.

Method: Construct a circle AB and two radii. Construct a parallel to each radius through the endpoint of the other.
Properties: A rhombus has equal consecutive sides and parallel opposite sides.

Method: Construct a segment AB and its midpoint C. Construct a line through point C, perpendicular to $\overline{AB}$. Construct circle CD, where point D is on the perpendicular line. Construct point E, the other intersection of the circle with the line. $ADBE$ is a rhombus.
Properties: The diagonals of a rhombus bisect each other.

Method: Construct a circle AB and then $\overline{BC}$, where point C is on the circle. Reflect point A across $\overline{BC}$. $ABA'C$ is a rhombus.
Properties: A rhombus has a pair of equal consecutive sides and a line of symmetry through their unshared endpoints.

Method: Construct a segment AB (to be a diagonal) and its midpoint, C. Construct a perpendicular through point C. Construct point D on the perpendicular. Reflect point D across $\overline{AB}$. $ADBD'$ is a rhombus.
Properties: The diagonals of a rhombus are perpendicular and are axes of reflection symmetry.

Method: Construct a segment *AB* to serve as half a diagonal. Construct a perpendicular through point *B*. Construct point *C* on the perpendicular. Rotate points *A* and *C* 180° about point *B*. *ACA′C′* is a rhombus.

Properties: The diagonals of a rhombus are perpendicular and their point of intersection is a center of 180° rotation symmetry.

Method: Construct a circle *AB* and point *C* on the circle. Bisect angle *BAC*. Construct circle *BA* and point *D*, the intersection of this circle and the bisector. *ABDC* is a rhombus.

Properties: The sides of a rhombus are equal and the diagonals bisect the angles.

Each of these methods has a corresponding example custom tool.

Constructing Rhombuses

How many ways can you come up with to construct a rhombus? Try methods that use the Construct menu, the Transform menu, or combinations of both. Consider how you might use diagonals. Write a brief description of each construction method along with the properties of rhombuses that make that method work.

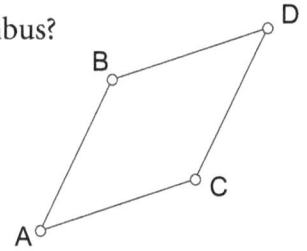

Method 1:

Properties:

_____

Method 2:

Properties:

_____

Method 3:

Properties:

_____

Method 4:

Properties:

The more time you give students, the more methods they'll come up with. This is probably the hardest of the special quadrilaterals to construct. Have students drag vertices of their figures to make sure their constructions are correct. Isosceles trapezoids that fall apart and can turn into other shapes are underconstrained. Isosceles trapezoids whose angles won't change are overconstrained. Here are various methods for constructing an isosceles trapezoid that is neither overconstrained nor underconstrained.

Method: Construct $\overline{AB}$ and $\overline{AC}$. Construct a line through point B parallel to $\overline{AC}$. Construct a circle with center C and radius AB. Find the point where the circle intersects the parallel line. This point is the fourth vertex of the trapezoid. (*Note:* Depending on how points are arranged, this construction may also be self-intersecting or be a parallelogram.)

Properties: An isosceles trapezoid has one pair of opposite sides that are equal and one pair that are parallel.

Method: Construct a horizontal line AB. Construct a ray AC using A as the endpoint. Mark angle CAB using the Transform menu. Mark point B as a center of rotation. Rotate line AB about point B by the marked angle. Construct a line through C parallel to line AB. Construct the intersection of this parallel line with the rotated line. This is the fourth vertex of the trapezoid. Hide all the lines and rays and connect the vertices with segments.

Properties: An isosceles trapezoid has one pair of opposite sides that are parallel and has congruent base angles.

Method: Construct a triangle ABC. Construct a line parallel to side AB through point C. Select point A and side BC and choose **Circle by Center+Radius.** Construct the point of intersection D of the circle and the parallel line. Hide the circle, the line, and segment BC. Construct the missing sides.

Properties: An isosceles trapezoid has one pair of opposite sides that are parallel and has congruent diagonals.

Method: Construct a circle *AB*. Construct a segment *AB*. Construct another radial segment *AC*. Construct segment *CB*. Construct *D*, a point on segment *AB*. Construct a line through *D* parallel to segment *CB*. Construct *E*, the point of intersection of segment *AC* and the parallel line. Hide parts of the construction that are not sides of the trapezoid, then construct the remaining sides of *DECB*.

Properties: An isosceles trapezoid has one pair of opposite sides that are parallel and one pair that are congruent.

Constructing Isosceles Trapezoids

How many ways can you come up with to construct an isosceles trapezoid? Try methods that use the Construct menu, the Transform menu, or combinations of both. Consider how you might use diagonals. Write a brief description of each construction method along with the properties of isosceles trapezoids that make that method work.

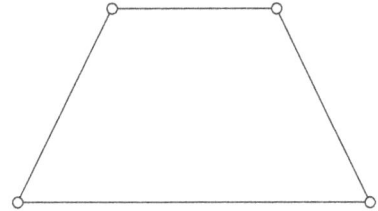

Method 1:

Properties:

_____

Method 2:

Properties:

_____

Method 3:

Properties:

_____

Method 4:

Properties:

Parabolas: A Geometric Approach

A good way to introduce this activity might be to talk a little about the history of parabolas. Parabolas have been studied for over 2000 years, but algebraic equations for them have only been known about for the past 500 years or so. How did people describe and define parabolas before? How were they drawn without coordinate geometry? Where do parabolas come from? (They are sections of cones created by a cutting plane parallel to the side of a cone). This kind of historical perspective may help students appreciate the true power of the algebraic equations.

CONSTRUCTION TIPS

2. Holding down the Shift key while drawing the line helps keep it horizontal.

3. Perform your first click the directrix itself (though not on one of its control points). (The directrix should be highlighted before you click.)

5. Use the **Arrow** tool in this step.

7. Refer to step 6 for a reminder on how to do this.

11. Make sure only point *F* is selected here. To deselect all objects, click in blank space.

12. Traces will gradually fade because of the Preference setting made in step 1. If you prefer that traces not fade, uncheck **Fade traces over time** on the Color panel of the Preferences dialog box. To clear traces from the screen, choose **Display | Erase Traces.**

13. Again, make sure only *F* is selected.

A PARABOLA FROM SCRATCH

Q1 It's the same distance from either endpoint. This is a theorem from geometry and can be proven, but for our purposes it's probably okay just to leave this up to students' common sense.

Q2 The two length measurements are always equal to each other. (This should confirm the students' conjectures from Q1.) The length of segment *FC* is the distance from *F* to the directrix, and the length of *FD* is the distance from *F* to the focus. Point *F*, then, fits the definition from the activity's introduction and is therefore on the parabola defined by the given focus and directrix.

Q3 The farther point C is from the directrix, the wider the parabola; the closer, the skinnier.

Q4 When point C is below the directrix, the parabola opens downward.

EXPLORE MORE

Q5 Select the focus and the directrix, and choose **Construct | Perpendicular Line.** Click the spot where this new line intersects the directrix to construct the point of intersection there. Hide the line and then construct a segment between the focus and the new point. Construct the midpoint of this segment using **Construct | Midpoint.** This point is the vertex of the parabola.

RELATED SKETCH

The page "Conic Sections" in **Parabolas Geometric Present.gsp** allows you to vary the angle at which a plane intersects a cone and to view the intersection from various directions.

Parabolas: A Geometric Approach

You may think of algebra and geometry as two very different branches of mathematics. In many ways they are. But you've seen that algebraic equations, such as $y = x + 3$, can be graphed as lines, which are geometric objects. Now you're studying parabolas—the graphs of equations such as $y = x^2 + 3$. Can parabolas be described geometrically, without using algebraic equations? In this activity you'll see that they can.

FOCUS AND DIRECTRIX

A *circle* can be described as the set of points in a plane that are the same distance from a fixed point—the center. Similarly, a *parabola* can be described as the set of points in a plane that are the same distance from a fixed point—the *focus*—as from a fixed line—the *directrix*.

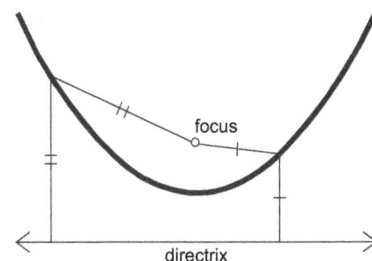

This is a more complicated definition than the circle's, but as you do this activity it should make more and more sense.

A PARABOLA FROM SCRATCH

If the **Line** tool isn't showing, press and hold on the current **Straightedge** tool and choose the **Line** tool from the palette that pops up.

1. In a new sketch choose **Edit | Preferences.** On the Text panel, check **Show labels automatically: For all new points.** On the Color panel, check **Fade traces over time.**

2. Using the **Line** tool, construct a horizontal line *AB.* (To make it exactly horizontal, hold the Shift key while you construct.) The line should be about a third of the way from the bottom of the sketch window. Show the line's label and change it to *directrix.*

3. Use the **Segment** tool to construct segment *CD* where *C* is on the line and *D* is about an inch above it.

4. Change point *D*'s label to *D: Focus.*

Click an object with the **Text** tool to show its label. Double-click the label itself to change it.

5. Construct the midpoint of segment *CD* by selecting it and choosing **Construct | Midpoint.**

6. Select segment *CD* and midpoint *E,* and choose **Construct | Perpendicular Line.**

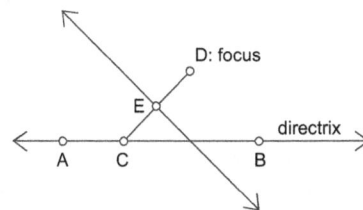

After step 6

The line you just constructed—perpendicular to segment *CD* through its midpoint—is called the *perpendicular bisector* of segment *CD.*

Q1 Imagine a point anywhere on the perpendicular bisector. How do you think the point's distance to *C* compares to its distance to *D*?

7. Construct a line perpendicular to $\overleftrightarrow{AB}$ and passing through point *C*.

8. Select the two perpendicular lines (the ones constructed in the last two steps) and choose **Construct | Intersection.**

You just constructed point *F*—the first point on the parabola.

To hide objects, select them and choose **Display | Hide.**

9. Hide $\overleftrightarrow{FC}$. Then, using the **Segment** tool, construct $\overline{FC}$ and $\overline{FD}$.

10. Measure the lengths of the two new segments. To do this, select them and choose **Measure | Length.**

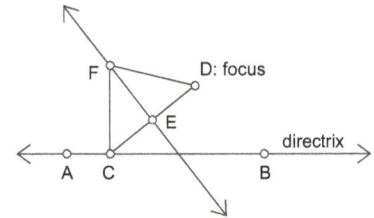

Q2 Using the **Arrow** tool, drag *C* back and forth along line *AB*.

After step 9

Refer back to the parabola definition in the introduction.

What do you notice about the length measurements as you drag *C*? Explain how this demonstrates that *F* is indeed always a point on the parabola defined by the given focus and directrix.

11. Select point *F* and choose **Display | Trace Intersection.** Now once again drag *C* back and forth. Point *F* will leave a trail as it traces out the parabola.

12. Drag the focus away from the directrix. Again, drag point *C*. Notice how this curve compares to the previous one. Now drag the focus closer to the directrix than it was originally and repeat the process.

Tracing the curve in this way works well enough to show you the shape of the parabola. But it can get a little annoying having to drag point *C* again and again. Here's a more efficient approach.

Select the point and choose **Display | Trace Intersection** to turn tracing on or off.

13. Turn off tracing for point *F*.

14. Select points *F* and *C*, and choose **Construct | Locus.**

The entire curve appears. It's called the *locus* of point *F* as point *C* moves along $\overleftrightarrow{AB}$.

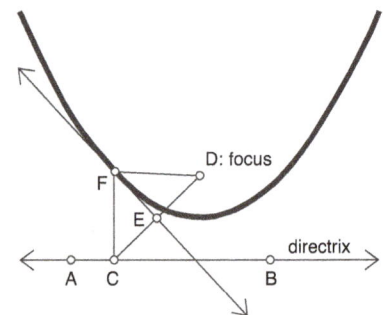

To hide objects, select them and choose **Display | Hide.**

15. To make the diagram less cluttered, hide everything except the parabola, the focus, and the directrix.

16. Drag the focus around and observe how the curve changes.

Q3 What happens to the parabola as the focus is dragged farther away from the directrix? Closer to it?

Q4 What happens to the parabola when the focus is dragged below the directrix?

17. Save your sketch. You may want to use it for another activity, Building Headlights and Satellite Dishes.

EXPLORE MORE

Q5 Use the commands in the Construct menu to construct the vertex of your parabola. This point should remain the vertex no matter where you drag the focus and the directrix.

Q6 Show that the equation of a parabola with its vertex at the origin and its focus at $(0, p)$ is $y = \frac{x^2}{4p}$. To do this, select the vertex you constructed in Q5 and choose **Define Origin** from the Graph menu. Then measure the y-coordinate (ordinate) of the focus and relabel this measurement p. Now use **Graph | Plot New Function** to plot $\frac{x^2}{4p}$. This plot should be right on top of the parabola, even if you drag the focus or the directrix.

DESCRIBE THE GEOMETRIC PARABOLA ALGEBRAICALLY

Q1 Here are some things that students may notice:

The *x*-coordinates of the focus and vertex are *h*.

The *y*-coordinate of the vertex is *k*.

The value of *k* is halfway between the *y*-coordinate of the focus and the *y*-value of the directrix.

In each case, the important thing is that students test their conjectures by adjusting the geometric parabola, fitting the algebraic one, and checking the results.

Q2 Encourage students to invent and test a variety of conjectures.

Q3 When the directrix is on the *x*-axis, the *y*-coordinate of the focus is twice the value of *k*.

Q4 When the focus is at the origin, the *y*-value of the directrix is twice the value of *k*.

13. There are several ways to do the construction. One is to construct the perpendicular from the focus to the directrix (the axis of symmetry). Then construct the midpoint of the segment from the focus to the intersection of the axis of symmetry and the directrix.

Q5 When the origin is halfway between the focus and directrix, $h = 0$ and the focus and directrix are equally distant from the origin.

Q6 When the directrix is not horizontal, you can no longer fit the graph to the geometric parabola.

EXPLORE MORE

Q7 The distance from the vertex to the focus is $\frac{1}{4a}$.

Q8 The coordinates of the focus are $(h, k + \frac{1}{4a})$. The equation of the directrix is $y = k - \frac{1}{4a}$.

From Locus to Graph

In this presentation you'll use algebraic methods to describe a parabola. Consider having a student operate the computer during the presentation.

A GEOMETRIC PARABOLA

If students are familiar with the definition and locus construction of a parabola, skip to step 3 to copy a geometric parabola.

1. Use the buttons on page 1 of **From Locus to Graph Present.gsp** to review the locus definition of a parabola.

You'll use page 2 to review the geometric construction of a parabola.

2. Use the first three buttons on page 2 to show the perpendicular bisector trace.

Q1 While the bisector is animating, ask, "How do you think the shape will change if we move the focus closer to the directrix?" Solicit several responses before dragging the focus. After dragging, ask students to explain why the shape changed in this way.

Point C is also constructed; this point will be the point of tangency of the perpendicular bisector and the parabola.

Q2 Ask students to explain how the shape will change if the focus is farther away.

3. Use the remaining buttons to construct the locus and select objects to copy. Then choose **Edit | Copy.**

AN ALGEBRAIC PARABOLA

We'd like to go beyond vague descriptions of shape by precisely analyzing the relationships between the shape of the parabola and the positions of focus and directrix. For that we will need analytic geometry—applying algebra to geometry. So let's make an algebraic parabola and analyze the connections.

*Choose **Graph | Plot New Function** and enter the equation.*

4. Page 3 contains three sliders, for a, h, and k. Use the values of the sliders to graph the function $f(x) = a(x - h)^2 + k$.

5. Drag a, h, and k to move the parabola around and change its shape.

6. Paste the previously copied geometric parabola by choosing **Edit | Paste.**

Q3 Ask, "How could we make the two parabolas match?" Solicit conjectures and test them by adjusting a, h, and k. Emphasize the value of incorrect conjectures in understanding how the parabola does and does not behave.

*To tabulate, select all these measurements and choose **Number | Tabulate.** Double-click the table to make the current entry permanent.*

7. Once the parabolas match, measure the equation of the directrix and the coordinates of the focus and vertex. Tabulate the values of a, h, k, and these three new measurements.

Q4 Ask, "What relationships can we find between the values of the sliders (the algebraic objects) and the measurements of the focus, directrix, and vertex (the geometric objects)?" Solicit and test a number of conjectures.

From Locus to Graph

Analytic geometry is the study of geometric shapes using the methods of algebra. Sometimes called *coordinate geometry*, it is fundamental to modern mathematics. René Descartes, who invented Cartesian coordinates, is considered the father of modern analytic geometry. In this activity you'll use algebraic methods to study and describe a parabola that you've defined geometrically.

A GEOMETRIC PARABOLA

Here's one geometric definition of a parabola:

A parabola is the set of points equidistant from a fixed point (the *focus*) and a fixed line (the *directrix*).

1. Begin with a geometric construction of a parabola, using either the activity Patty Paper Parabolas or page 2 of **From Locus to Graph.gsp.**

AN ALGEBRAIC PARABOLA

Here's one algebraic formulation of a parabola:

An equation in the form $f(x) = a(x - h)^2 + k$ has as its graph a parabola with a horizontal directrix.

2. Graph this equation using sliders for a, h, and k, and determine the role that each of these values plays in determining the shape of the graph. (The activity Parabolas in Vertex Form describes how to do this, or you can use page 3 of **From Locus to Graph.gsp.**)

DESCRIBE THE GEOMETRIC PARABOLA ALGEBRAICALLY

3. In the sketch containing the geometric parabola, make the directrix precisely horizontal.

To copy the objects, select them all and choose **Edit | Copy.**

4. Copy the focus, the directrix, and the locus (the parabola itself).

5. Paste the geometric parabola into the sketch containing the algebraic graph of a parabola (or into page 3 of the sketch **From Locus to Graph.gsp**).

6. Adjust the sliders that control the values of a, h, and k until the algebraic parabola exactly matches the geometric one. Notice the values of a, h, and k.

You can measure the equation of a line, but not a segment. If the directrix is a segment, construct a line on top of it and measure the equation of the line.

7. Measure the equation of the directrix, and measure the coordinates of the focus.

Q1 What connections do you think there might be between these measurements and the values of the sliders? Write down one conjecture and then use the next few steps to test it.

8. Tabulate the two measurements from step 7 and the values of a, h, and k. (Do this by selecting the two measurements and the values a, h, and k, and then choosing **Number | Tabulate.**)

9. Double-click the table to make the current values permanent.

10. Move the directrix and focus to new locations. Be sure to keep the directrix horizontal as you move it.

11. Adjust the sliders until the algebraic parabola again matches the geometric one. Double-click the table to make these values permanent.

12. Repeat steps 10 and 11 several more times, experimenting with different positions for the focus and directrix.

Q2 Write down several more conjectures about the connections between the algebraic values (a, h, and k) and the measurements of the geometric objects. Test your conjectures by using the data already in the table, or by collecting additional data.

Q3 What happens when the directrix is on the x-axis? Why?

Q4 What happens when the focus is at the origin? Why?

13. The vertex of the parabola is halfway between the focus and directrix. Using these two objects, construct the vertex geometrically and measure its coordinates.

Q5 What happens when you place the vertex at the origin? Why?

Q6 What happens when the directrix is not horizontal? Explain.

EXPLORE MORE

14. Experiment to find how a is related to the distance from the directrix to the vertex. Measure this distance, and then make a new table containing these two values (a and the distance you just measured). Place the directrix on the x-axis. Then move the focus to $(0, 1)$ and add these values to the table. Do this again with the focus at different positions, such as $(0, 2)$, $(0, 3)$, $(0, 0.5)$, and $(0, -1)$.

Q7 What can you conclude about the relationship between a and the distance from the vertex to the directrix?

Q8 Express the coordinates of the focus and the equation of the directrix in terms of a, h, and k.

If it's not convenient to take such a photo, find one on the web.

15. Use a digital camera to photograph a curved shape that seems to be parabolic. Copy the picture and paste it into your sketch containing the algebraic graph of a parabola. Adjust a, h, and k to see if you can fit the graph to the picture.

7

Triangle Properties (Geometry)

Triangle Congruence

For GSP5 ACTIVITY NOTES

SKETCH AND INVESTIGATE

Q1 No. Three sides determine a unique triangle.

Q2 If two triangles have three pairs of congruent sides, then the triangles are congruent (SSS).

Q3 These combinations of parts guarantee congruence: SSS, SAS, ASA, AAS. For AAS, it's important to state that the sides must correspond. (The correspondence is forced by the order in the cases: It's impossible for the parts not to correspond.)

SSA and AAA do not guarantee congruence.

EXPLORE MORE

1. Counterexample for SSA:

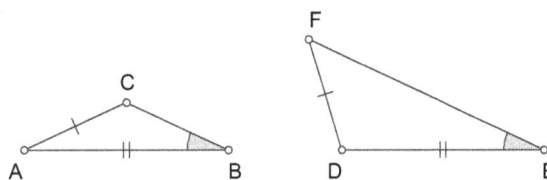

Counterexample for AAA:

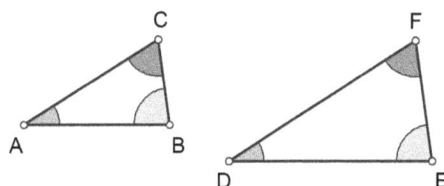

Triangle Congruence

If the three sides of one triangle are congruent to three sides of another triangle (SSS), must the two triangles be congruent? What if two sides and the angle between them in one triangle are congruent to two sides and the angle between them in another triangle (SAS)? Which combinations of parts guarantee congruence and which don't? In this activity you'll investigate that question.

SKETCH AND INVESTIGATE

1. Open the sketch **Triangle Congruence.gsp**. You'll see a figure like the one shown below, along with some text.

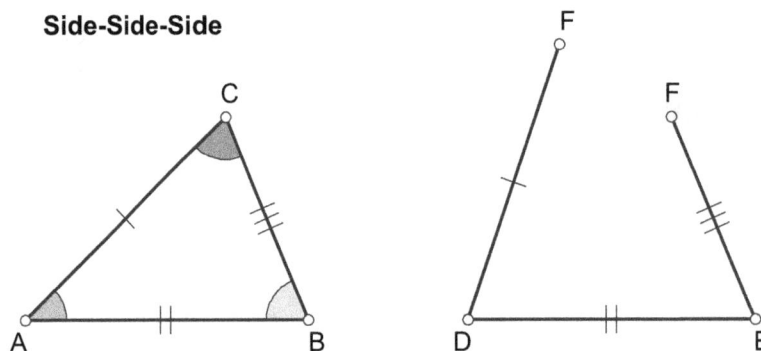

Side-Side-Side

2. Read the text in the sketch and follow the instructions to try to make a triangle DEF that is not congruent to △ABC.

Q1 Could you form a triangle with a different size or shape given the three sides? _____

Q2 If three sides of one triangle are congruent to three sides of another (SSS), what can you say about the triangles? Write a conjecture that summarizes your findings.

3. Go to the SAS page and make a triangle *DEF* with those given parts. Try to make a triangle that is not congruent to △ABC.

4. Experiment on each of the other pages.

Q3 Of SSS, SAS, SSA, ASA, AAS, and AAA, which combinations of corresponding parts guarantee congruence in a pair of triangles? Which do not?

> Click on the page tabs at the bottom of the sketch window to switch pages.

EXPLORE MORE

1. For each of the combinations of parts that do not necessarily guarantee congruence, sketch a pair of noncongruent triangles with those congruent parts. Explain why the triangles are not congruent.

Similar Triangles—AA Similarity

SKETCH AND INVESTIGATE

Q1 Because $\angle D \cong \angle B$ and $\angle E \cong \angle C$, angles A and F are congruent. This is because they contain the degrees that remain after you subtract m$\angle B$ and m$\angle C$ from 180°.

Q2 Triangles with two pairs of corresponding angles congruent must be similar.

EXPLORE MORE

1. Neither AA nor AAA is a sufficient condition for similarity in quadrilaterals. A square and a rectangle provide a simple counterexample because they have all four corresponding angles congruent, but they are not necessarily similar. Students can prepare Sketchpad constructions similar to the one in this activity to demonstrate this.

2. The image below shows the construction. Triangle *LKI* and triangle *LJC* are similar. The pairs of angles are congruent as marked because each pair is a pair of corresponding angles. The triangles are therefore similar by AA.

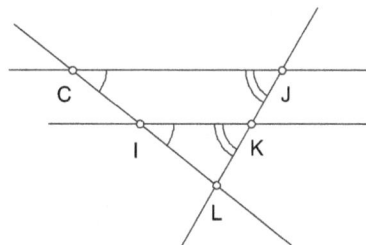

Similar Triangles—AA Similarity

To show that two triangles are similar, you don't need to show that all the corresponding angles are congruent and that all the corresponding sides are proportional. There are several shortcuts. In this activity you'll construct and investigate triangles with two pairs of congruent corresponding angles.

SKETCH AND INVESTIGATE

1. Construct △ABC.

Double-click point D to mark it as a center. Select, in order, points C, B, and A; then, in the Transform menu, choose **Mark Angle.**

2. Construct $\overleftrightarrow{DE}$ below the triangle.

3. Mark point D as a center of rotation and mark ∠CBA as an angle of rotation.

Select $\overleftrightarrow{DE}$; then, in the Transform menu, choose Rotate.

4. Rotate $\overleftrightarrow{DE}$ by the marked angle.

5. Drag point A and observe the effect on the angle formed by the rotated line.

6. Mark E as a center and mark ∠BCA.

7. Rotate $\overleftrightarrow{DE}$ about the new marked center by the new marked angle.

8. Construct point F, the point of intersection of these two rotated lines.

9. Hide the lines and replace them with segments.

10. Drag vertices of △ABC and observe the effects on △FDE.

Q1 You constructed angles D and E to be congruent to angles B and C, respectively. Without measuring, how do you think angles A and F compare? Explain why they're related in this way.

Select two corresponding segments; then, in the Measure menu, choose **Ratio.** Repeat for the other two pairs of sides.

11. To see if the triangles are similar, measure the ratios of all three pairs of corresponding side lengths. Drag points and observe these ratios.

Q2 Make a conjecture about triangles with two pairs of congruent corresponding angles.

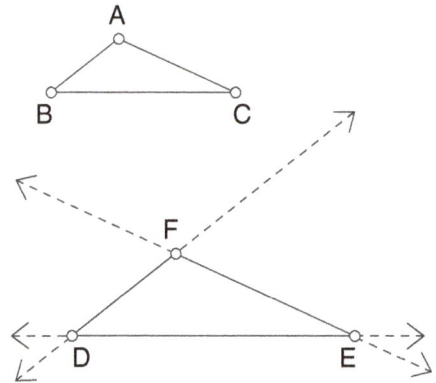

EXPLORE MORE

1. Investigate AA or AAA similarity for quadrilaterals

2. Construct two parallel lines and two nonparallel transversals. The transversals should intersect to form similar traingles. How do you know these triangles are similar?

Similar Triangles—SSS, SAS, SSA

SKETCH AND INVESTIGATE

Q1 If two triangles have all three corresponding pairs of sides proportional, the triangles are similar.

Q2 The SSS and SAS combinations of corresponding parts guarantee similarity in a pair of triangles. The combination SSA does not. In the diagram below, $ED/BA = DF/AC$ and m$\angle F$ = m$\angle C$. However, the triangles are clearly not similar.

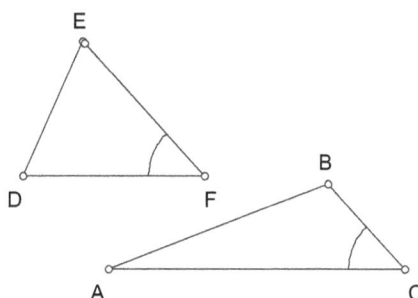

EXPLORE MORE

4. ASASA and SASAS, for example, are enough to determine similarity in quadrilaterals. No combinations of four-part pairings guarantee similarity, but there are other five-part combinations that do.

Similar Triangles—SSS, SAS, SSA

You may have already discovered that if two angles in one triangle are congruent to two angles in another triangle, the triangles are similar. In this activity you'll experiment with a sketch to discover other shortcuts for determining whether two triangles are similar. These shortcuts involve different combinations of proportional sides and congruent angles.

SKETCH AND INVESTIGATE

1. Open the sketch **Triangle Similarity.gsp.** You'll see a sketch like the one shown below.

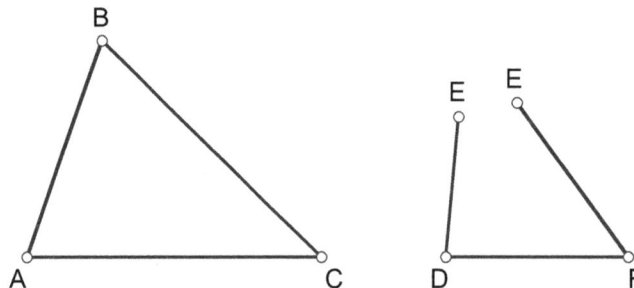

Press any of the buttons below to try that triangle similarity experiment.

| SSS | SAS | SSA |

2. Press the SSS action button and follow the instructions on the screen.

Q1 Write your conjecture below.

3. Experiment with the SAS button and the SSA button.

Q2 Of SSS, SAS, and SSA, which combinations of corresponding parts guarantee similarity in a pair of triangles? Which do not?

EXPLORE MORE

4. Investigate similarity in quadrilaterals. What's the smallest amount of information you need to determine that quadrilaterals are similar?

Constructing Squares on a Triangle:
The Pythagorean Theorem

For
GSP5

ACTIVITY NOTES

SKETCH AND INVESTIGATE

Q1 This construction uses the following property: A square has right angles and congruent sides.

Q2 This construction uses the following property: A right triangle has one side perpendicular to another side.

Q3 The sum of the areas of the two smaller squares equals the area of the largest square. If the measurements are displayed with precision greater than the tenths place, students may not notice the relationship. It might help them to choose **Edit | Preferences | Units** and change their Distance Precision to **tenths.**

Q4 $a^2 + b^2 = c^2$

Students already familiar with the Pythagorean theorem are likely to write this as their conjecture. You might also have them express the theorem in words.

EXPLORE MORE

22. The areas of the regions built on the legs of a right triangle will always sum to the area of the region built on the hypotenuse, as long as all three regions are similar. If students have custom tools handy, they can check this easily: Try regular hexagons on each of the three sides or equilateral triangles.

23. If the areas of squares on two sides of a triangle sum to the area of the square on the third side, the triangle must be a right triangle.

Constructing Squares on a Triangle: The Pythagorean Theorem

For GSP3

In this investigation you'll create a custom tool for constructing a square, and then you'll construct squares on the sides of a right triangle. The areas of these squares illustrate perhaps the most famous relationship in mathematics—the Pythagorean theorem.

SKETCH AND INVESTIGATE

1. Construct $\overline{AB}$.

> Double-click point *A* to mark it as a center. Select point *B* and $\overline{AB}$; then, in the Transform menu, choose **Rotate**.

2. Mark point *A* as a center and rotate point *B* and $\overline{AB}$ by 90°.

3. Mark point *B′* as a center and rotate point *A* and $\overline{B'A}$ by 90°.

4. Construct $\overline{A'B}$ to finish the square.

> Select the vertices in consecutive order; then, in the Construct menu, choose **Quadrilateral Interior**.

5. Construct the interior.

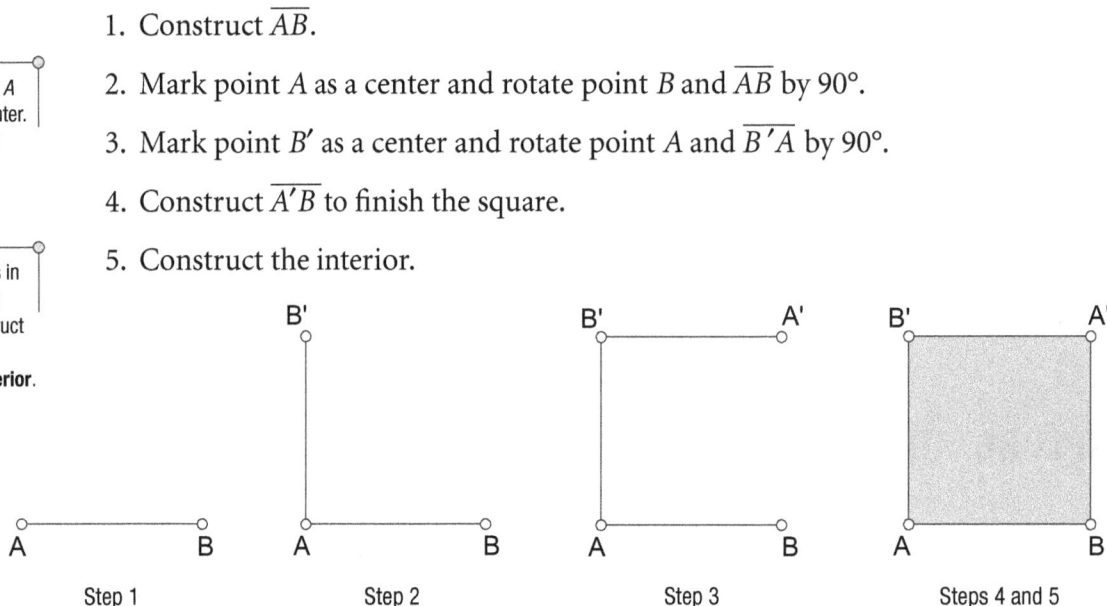

Step 1   Step 2   Step 3   Steps 4 and 5

> Use the **Text** tool and click each point.

6. Drag each vertex of the square to make sure it holds together.

7. Hide the labels.

> Select the entire figure; then, in the Custom Tools menu (the bottom tool in the Toolbox), choose **Create New Tool**.

8. Make a custom tool of this construction.

Q1 What properties of a square did you use in this construction?

> You might want to save this tool so that you can use it in the future. See the appropriate sections in the help system. (Choose **Tools** from the Help menu, and then click the Custom Tools link.)

9. Experiment with using the custom tool to get a feel for the way it works. Note that the direction in which the square is constructed depends on how you use the tool.

10. Open a new sketch.

11. Construct $\overline{AB}$.

> Select point *A* and $\overline{AB}$ and, in the Construct menu, choose **Perpendicular Line.**

12. Construct a line through point *A* perpendicular to $\overline{AB}$.

13. Construct $\overline{BC}$, where point *C* is a point on the perpendicular line.

14. Hide the perpendicular line and construct $\overline{AC}$.

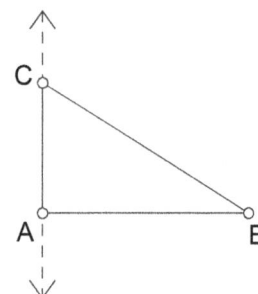

15. Drag each vertex to confirm that your triangle stays a right triangle.

Q2 What property of a right triangle did you use in your construction?

16. Change labels so that the right-angle vertex is labeled *C* and the other two vertices are labeled *A* and *B*.

To change a label, double-click the label with the finger of the **Text** tool. (The reason for changing these labels is so that your figure will match the way the theorem is usually stated. This may make it easier to remember the theorem.)

17. Show the labels of the sides. Change them to *a*, *b*, and *c* so that side *a* is opposite ∠*A*, side *b* is opposite ∠*B*, and side *c* is opposite ∠*C*.

18. Use your square tool to construct squares on the sides of your triangle.

Be sure to attach each square to a pair of the triangle's vertices. If your square goes the wrong way (overlaps the interior of your triangle) or is not attached properly, undo and try attaching the square to the triangle's vertices in the opposite order.

19. Drag the vertices of the triangle to make sure the squares are properly attached.

20. Measure the areas of the three squares.

21. Measure the lengths of sides *a*, *b*, and *c*.

22. Drag each vertex of the triangle and observe the measures.

Choose **Calculate** from the Number menu to open the Calculator. Click a measurement to enter it into a calculation.

Q3 Describe any relationship you see among the three areas. Use the Calculator to create an expression that confirms your observations.

Q4 Based on your observations about the areas of the squares, write an equation that relates *a*, *b*, and *c* in any right triangle. (*Hint:* What's the area of the square with side length *a*? What are the areas of the squares with side lengths *b* and *c*? How are these areas related?)

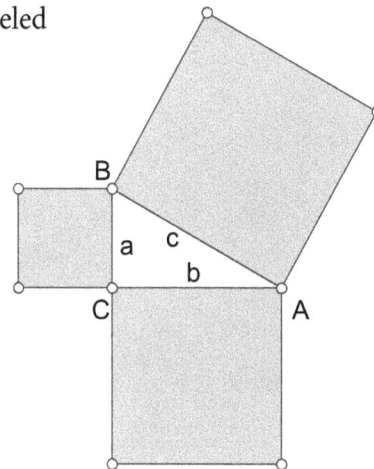

EXPLORE MORE

22. Do a similar investigation using other figures besides squares. Does your conjecture about the areas still hold?

23. Investigate the converse of the Pythagorean theorem: Construct a nonright triangle and squares on its sides. Measure the areas of the squares and sum two of them. Drag until the sum is equal to the third area. What kind of triangle do you have?

The Isosceles Right Triangle

SKETCH AND INVESTIGATE

Q1 Triangle *ABC* is right and isosceles because two of its sides are sides of a square. Thus, they're congruent and they form a right angle.

Q2 Each acute angle in $\triangle ABC$ measures 45°.

Q3 The ratio is constant at 1.414. (Decimals will vary depending on the precision students have set in Preferences.)

Q4 If each of the smaller squares has an area of x^2, the area of the large square is $2x^2$. (Each small square can be divided into two isosceles right triangles, four of which fit into the large square.)

Q5 If the legs of the isosceles triangle have length x, the length of the hypotenuse is $x\sqrt{2}$. This is because the area of the square on the hypotenuse is $2x^2$, so the length of a side of the square must be, $\sqrt{2x^2}$ or $x\sqrt{2}$.

Q6 According to Q5, if each leg has length x, the hypotenuse has length $x\sqrt{2}$. Substituting these values in the Pythagorean theorem gives $x^2 + x^2 = (x\sqrt{2})^2 = 2x^2$.

EXPLORE MORE

10. The discovery of the irrationality of $\sqrt{2}$ originated with the isosceles right triangle with sides of length 1. The Pythagoreans were startled to find a number that could not be represented as the ratio of two integers. The standard proof of the irrationality of $\sqrt{2}$ is by contradiction: assume the number is rational and show that this leads to impossible conclusions. The tenth book of Euclid's *Elements* contains a discussion similar to this proof. And a reference in one of Aristotle's works makes it clear that the proof was known much earlier than Euclid.

The Isosceles Right Triangle

In this activity you'll discover a relationship among the side lengths of an isosceles right triangle. This relationship will give you a shortcut for finding side lengths quickly. You'll start by constructing a square. Dividing this square in half along a diagonal gives you the isosceles right triangle.

SKETCH AND INVESTIGATE

You can use the tool **4/Square (By Edge)** from the sketch **Polygons.gsp**.

1. Use a custom tool to construct a square *ABCD* by edge endpoints *A* and *B*.

2. Construct diagonal *CA*.

Hide and **Line Style** are in the Display menu.

3. Hide the square's interior, if it has one.

4. Change the line styles of $\overline{CD}$ and $\overline{DA}$ to dashed.

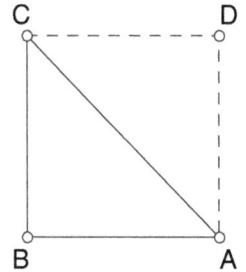

Q1 Explain why △*ABC* is an isosceles right triangle.

Q2 Without measuring, state the measures of the acute angles in △*ABC*.

5. Measure the three side lengths.

Select the hypotenuse and one of the legs. Then, in the Measure menu, choose **Ratio.**

6. Measure the ratio of the hypotenuse length to one of the leg lengths.

7. Drag point *A* or point *B* and observe this ratio.

Q3 What do you notice about this ratio?

In steps 8 and 9, you'll investigate what this ratio represents geometrically.

8. Use the square tool to construct squares on the sides of right triangle *ABC*. Drag to make sure the squares are properly attached.

9. Construct one diagonal in each of the smaller squares, as shown at right.

Q4 The diagonals you drew in the smaller squares may help you see a relationship between the smaller squares and the square on the hypotenuse. If each of the smaller squares has area x^2, what is the area of the large square? Confirm your conjecture by measuring the areas. Drag the triangle to confirm that this relationship always holds true.

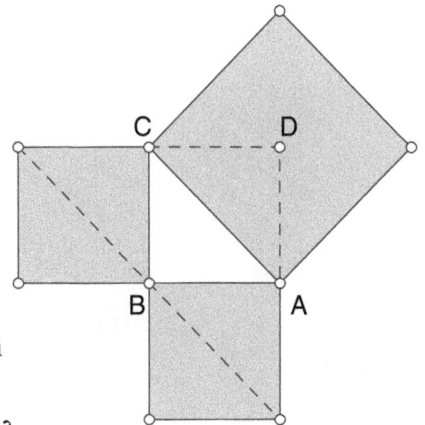

Q5 In an isosceles right triangle, if the legs have length x, what is the length of the hypotenuse?

Q6 Use the Pythagorean theorem to confirm your answer to Q5.

EXPLORE MORE

10. The society of the Pythagoreans discovered that the square root of 2 is *irrational*. Do some research and report on this discovery.

SKETCH AND INVESTIGATE

Q1 $\angle CDB$ measures 90° because the median of an equilateral triangle is also an altitude. $\angle DBC$ measures 60° because it is an angle in an equilateral triangle. $\angle BCD$ measures 30° because it is half of an angle in an equilateral triangle.

Q2 In a 30°-60° right triangle, the hypotenuse is twice as long as the shortest leg.

Q3 The ratio of the area of the largest square to the area of the smallest square is 4:1. The areas have a ratio of 4:1 because the sides have a ratio of 2:1 and $2^2:1^2 = 4:1$.

Q4 a. If the smallest square has area x^2, the largest square will have area $4x^2$.

b. If A represents the area of the square on the long leg, the Pythagorean theorem gives us $x^2 + A = 4x^2$. Subtracting x^2 from both sides gives us $A = 3x^2$.

c. The ratio of the area of the square on the long leg to the area of the square on the short leg is 3:1. (If you can, check to make sure students verify this in their sketch.)

Q5 a. If the short leg has length x, the hypotenuse has length $2x$.

b. The long leg would have length $x\sqrt{3}$ because the leg is the side of the square with area $3x^2$.

c. The ratio $x\sqrt{3} : x$ simplifies to $\sqrt{3}:1$.

Q6 1.713 (Answers will vary depending on the Precision setting for calculations. Change this in Preferences in the Edit menu.)

EXPLORE MORE

13. To make a Hide/Show button, select the object you want to show and hide; then choose **Edit | Action Buttons | Hide/Show**. Double-click any button with the **Text** tool to change its label. A student who finishes the activity early might make this sketch. Then you can use it with the whole class to practice calculating the missing sides for different 30°-60° right triangles.

The 30°-60° Right Triangle

The 30°-60° right triangle—formed by taking half of an equilateral triangle—has special relationships among its side lengths. These relationships make it easy to find all the side lengths if you know just one. In this activity you'll discover these relationships and why they hold.

SKETCH AND INVESTIGATE

You can use the tool **3/Triangle (By Edge)** from the sketch **Polygons.gsp.** If you need to relabel the triangle, use the **Text** tool. Click a point to show its label. Double-click a label to change it.

1. In a new sketch, construct an equilateral triangle *ABC*. Use a custom tool or construct it from scratch. Drag vertices to confirm that the construction is correct.

2. Hide the triangle's interior if necessary.

3. Construct the midpoint *D* of $\overline{AB}$.

4. Construct median *CD*.

Select the segments; then, in the Display menu, choose **Line Style | Dashed.**

5. Change the line styles of $\overline{AB}$ and $\overline{AC}$ to dashed.

6. Construct $\overline{BD}$ and make its line style thin.

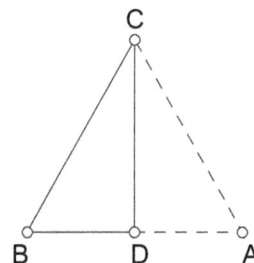

Q1 Without measuring, state the measure of each angle in △*CDB*. For each angle, explain how you know it has that measure.

Q2 In a 30°-60° right triangle, how does the length of the hypotenuse compare to the length of the short leg? Answer without measuring.

7. Hide point *A*, $\overline{AB}$, and $\overline{AC}$.

8. Use a custom tool to construct squares on the three sides of the triangle as shown at right. Drag to make sure the squares are properly attached.

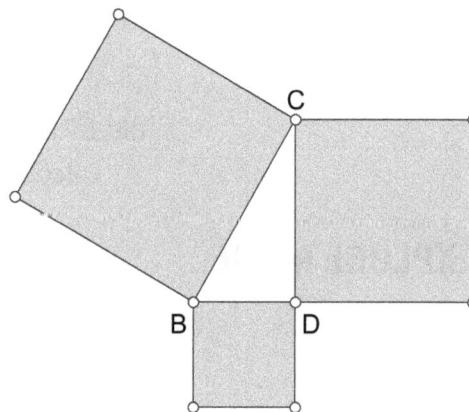

9. Measure the areas of the three squares.

Choose **Calculate** from the Number menu to open the Calculator. Click a measurement to enter it into a calculation.

10. Calculate the ratio of the largest area to the smallest area

11. Drag point *B* and observe this ratio.

Q3 What is this area ratio? Explain why this ratio is what it is. In your explanation, use what you know about the side lengths.

Q4 Now you'll use your answer to Q3 about the square on the hypotenuse and the square on the short leg to help you find the area of the square on the long leg. Answer the following questions.

    a. Suppose the smallest square has area x^2. What would be the area of the square on the hypotenuse? _____

    b. Use the Pythagorean theorem to find the area of the square on the long leg. Show your work.

    c. State the ratio of the area of the square on the long leg to the area of the square on the short leg. Calculate this ratio in your sketch and drag point B to confirm that this ratio applies to all 30°-60° right triangles. _____

Q5 Suppose the short leg had length x.

    a. What would be the length of the hypotenuse? _____

    b. What would be the length of the long leg? _____

    c. What would be the ratio of the length of the long leg to the length of the short leg? (State your answer in radical form.) _____

Select the hypotenuse and the short leg. Then, in the Measure menu, choose **Ratio**. Measure the other ratio in the same way.

12. To confirm your answers to Q5, measure the ratio of the hypotenuse length to the short leg length. Also measure the ratio of the long leg length to the short leg length. Drag point B to confirm that these ratios apply to all 30°-60° right triangles.

Q6 The second ratio you measured in step 12 should be the decimal approximation of the ratio you wrote in Q5c. Record this decimal approximation. _____

EXPLORE MORE

13. To test how well you can apply your discoveries, make a Hide/Show action button for each side length measurement. Show one side length and hide the other two. Then calculate the two hidden lengths. Show the hidden lengths to check your calculations. Try this several times, changing the triangle each time and showing different side lengths. Repeat until you think you can calculate the missing side lengths correctly every time.

Circles and Transformations (Geometry)

Chords in a Circle

SKETCH AND INVESTIGATE

Q1 The perpendicular bisector of any chord of a circle passes through the center of the circle.

Q2 The closer the chord is to the center of the circle, the longer the chord. The chord is longest when its distance to the center is zero. This chord is a diameter of the circle.

Q3 At the point where the locus intersects the y-axis, the length of $\overline{BC}$ is zero (its minimum value) and the length of $\overline{AD}$ is the radius of the circle (its maximum value). Likewise, at the point where the locus intersects the x-axis, the length of $\overline{BC}$ is the diameter of the circle (its maximum value) and the length of $\overline{AD}$ is zero (its minimum value).

As point G moves from left to right, its y-coordinate decreases in value, showing the chord $\overline{BC}$ moving closer to the center of the circle (and also becoming longer).

Students may notice that the locus shows a portion of an ellipse in the first quadrant. This ellipse is centered at the origin and has a major axis of twice the diameter of the circle and a minor axis of twice the radius of the circle.

Q4 If two chords in a circle are congruent, the chords are the same distance from the center of the circle. (The converse is also true.)

EXPLORE MORE

1. When $HI = BC$, the plotted point (length of $\overline{HI}$, distance from $\overline{HI}$ to the center) is coincident with point G.

2. After constructing the arc, construct two chords anywhere on the arc. Construct the perpendicular bisectors of both these chords. Their point of intersection is the center of the circle.

Chords in a Circle

A chord in a circle is a segment with endpoints on the circle. In this activity you'll investigate properties of chords.

SKETCH AND INVESTIGATE

1. Construct circle AB.

2. Construct chord BC.

3. Construct the midpoint D of the chord.

Select point D and $\overline{BC}$; then, in the Construct menu, choose **Perpendicular Line.**

4. Construct a line through point D perpendicular to $\overline{BC}$. This line is the perpendicular bisector of the chord. (It's not shown in the figure.)

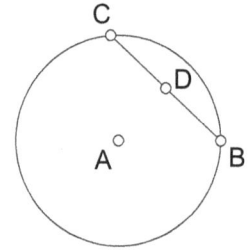

5. Drag point C around the circle and observe the perpendicular line.

Q1 Write a conjecture about the perpendicular bisector of any chord in a circle.

Select the line; then, in the Display menu, choose **Hide Line.**

6. Hide the perpendicular bisector and construct $\overline{AD}$.

7. Measure the length of $\overline{AD}$. This is the distance from the chord to the center.

8. Measure the length of $\overline{BC}$.

9. Drag point C around the circle and observe the measures.

Q2 How is the length of the chord related to its distance from the center?

If you don't see point G, scale the axes by dragging point F toward point E.

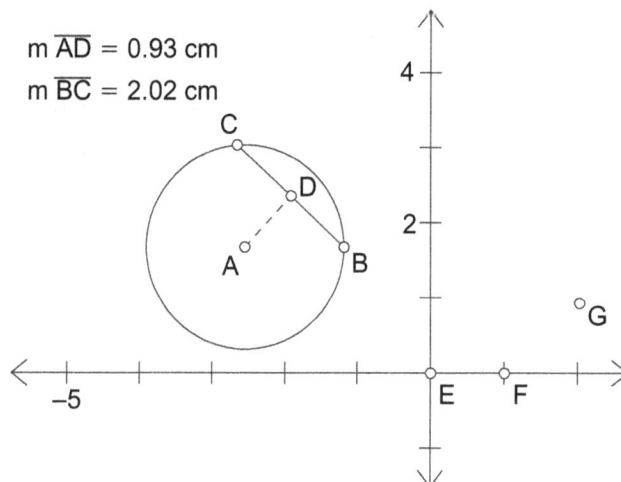

10. You can make a graph that shows this relationship: Select the length of $\overline{BC}$ and the length of $\overline{AD}$, in that order; then, in the Graph menu, choose **Plot As (x, y)**. You should get axes and a point G whose coordinates are the measures you selected.

m $\overline{AD}$ = 0.93 cm
m $\overline{BC}$ = 2.02 cm

11. Drag point *C* to see how it controls point *G*.

12. To graph all the possible locations for point *G*, select it and point *C*; then, in the Construct menu, choose **Locus**.

13. Drag point *C* to see point *G* travel along the locus.

14. Drag point *A* or point *B* to see what effect changing the circle's radius has on the graph.

Q3 Write a paragraph describing the graph. Answer these questions in your paragraph: Look at the value of *y* where the locus intersects the *y*-axis. What does this value represent in the circle? Look at the value of *x* where the locus intersects the *x*-axis. What does this value represent in the circle? As point *G* moves from left to right, what happens to the value of its *y*-coordinate? What does this have to do with what's happening to the chord? Use a separate sheet of paper.

15. Construct $\overline{HI}$, another chord on the circle.

16. Measure *HI*.

Select $\overline{HI}$ and point *A*; then, in the Measure menu, choose **Distance.**

17. Measure the distance from $\overline{HI}$ to the center of the circle.

18. Drag point *H* or point *I* and watch the length measure. Try to make this length as close to the length of $\overline{BC}$ as you can.

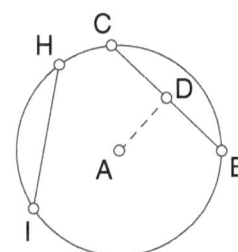

Q4 Write a conjecture about congruent chords in a circle.

EXPLORE MORE

1. Plot as (*x*, *y*) the length of $\overline{HI}$ and the distance from $\overline{HI}$ to the center. How does this plotted point compare to point *G* when *HI* = *BC*?

2. An arc is part of a circle. You can construct an arc from any three points. In a new sketch, construct a three-point arc. Now use your conjecture from Q1 to construct the center of the circle containing the arc. Construct the circle to confirm that you found the correct point. Explain what you did.

SKETCH AND INVESTIGATE

Q1 As you drag point *C* toward point *B*, the angle between the radius and the secant approaches 90°. If you drag slowly, you can tell when points *C* and *B* coincide, the secant becomes a tangent, and ∠*ABC* measures 90°.

Q2 A tangent is perpendicular to the radius at the point of tangency.

Q3 Select the radius and its endpoint on the circle, then, in the Construct menu, choose **Perpendicular Line.**

EXPLORE MORE

6. To construct an internally or externally tangent circle, first construct a circle and a tangent line. Extend the radius to a full line and construct a second circle with its center on that line and its radius endpoint the same as that of the first circle (the point of tangency).

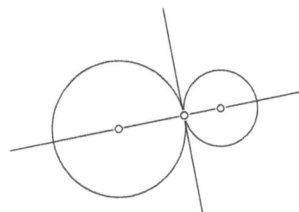

Tangents to a Circle

A line can intersect a circle in zero, one, or two points. A line that intersects a circle in exactly one point—that just touches the circle without going into the circle's interior—is called a *tangent*. The point of intersection is called the *point of tangency*. A line that intersects a circle in two points is called a *secant*. In this investigation you'll construct a secant, then manipulate it until it becomes a tangent to discover an important property of tangents.

SKETCH AND INVESTIGATE

1. Construct circle *AB*.

2. Construct $\overline{AB}$.

Press and hold the pointer on the **Segment** tool; then drag right to choose the **Line** tool.

3. Construct secant $\overline{BC}$, making sure point *C* falls on the circle.

4. Measure ∠*ABC*.

Select, in order, points *A, B,* and *C.* Then, in the Measure menu, choose **Angle.**

5. Drag point *C* around the circle and observe the angle measure.

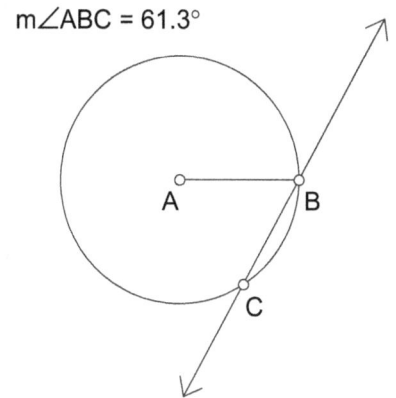

m∠ABC = 61.3°

Q1 What happens to m∠*ABC* as point *C* gets closer to point *B*? What's the measure of ∠*ABC* when point *C* is right on top of point *B*?

Q2 When points *B* and *C* coincide, your line intersects the circle in a single point, so it's tangent to the circle. How is a tangent related to the radius at the point of tangency?

Q3 Use what you observed in Q2 to construct a line in your sketch that is always tangent to the circle. Describe how you did it.

EXPLORE MORE

6. Come up with methods for constructing two circles that always intersect in one point. The circles could be *internally tangent* (one inside the other) or *externally tangent* (neither inside the other).

Tangent Segments

SKETCH AND INVESTIGATE

Q1 Tangent segments to a circle from a point outside the circle are congruent.

EXPLORE MORE

1. The right triangles *ABD* and *ACD* are congruent by Hypotenuse-Leg. (*AB* and *AC* are radii of a circle and $\overline{AD}$ is congruent to itself.) Thus, $\overline{DB}$ and $\overline{DC}$ are corresponding parts of congruent triangles and must be congruent.

2. This is a tough challenge. The diagram below shows the necessary construction. Circle *A* is the original circle and point *C* is the point outside the circle. Point *D* is the midpoint of $\overline{AC}$. The lines are perpendicular to the radial segments because they create angles inscribed in semicircles.

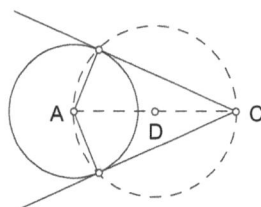

3. Construct a circle and a tangent to the circle. Construct another point *P* on the tangent and construct a perpendicular line through point *P*. Then construct a circle centered on the perpendicular with point *P* its other control point.

Tangent Segments

In this activity you'll learn how to construct tangents. Then you'll compare the lengths of two tangent segments from the common intersection point to the points of tangency.

SKETCH AND INVESTIGATE

1. Construct circle *AB* and radius *AB*.

Select $\overline{AB}$ and point *B*; then, in the Construct menu, choose **Perpendicular Line.**

2. Construct a line perpendicular to $\overline{AB}$ through point *B*. This line is tangent to the circle.

3. Drag point *B* to confirm that the line stays tangent.

4. Construct a second radius *AC*.

5. Construct a tangent through point *C*.

6. Drag point *C* to confirm that this line stays tangent.

7. Construct point *D* where the tangent lines intersect.

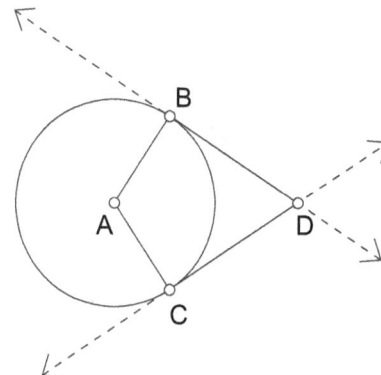

Select the lines; then, in the Display menu, choose **Hide Lines.**

8. Hide the lines.

9. Construct segments *BD* and *CD*.

10. Measure *BD* and *CD*.

11. Drag point *C* and observe the measures.

Q1 Write a conjecture about tangent segments.

EXPLORE MORE

1. Construct $\overline{AD}$. Investigate relationships among the angles and sides of the two triangles formed. Are the triangles congruent? If so, explain why.

2. Construct a circle and a point outside the circle. Come up with a method for constructing two tangents from the given point. Describe your method.

3. Come up with methods for constructing two or more circles with a common tangent (a line tangent to both circles). *Hint:* Construct the second circle after you've constructed the first circle's tangent. Describe your method.

Arcs and Angles

SKETCH AND INVESTIGATE

Q1 The measure of a central angle in a circle equals the measure of the minor arc it intercepts.

Q2 The measures of the central angle and the major arc it intercepts sum to 360°. (The measure of the major arc is equal to 360° minus the measure of the central angle.)

Q3 The measure of an inscribed angle is always half the measure of the arc it intercepts. (Notice that if you drag *C* beyond *D*, the arc measurement no longer shows the measure of the arc intercepted by ∠ *CDB*.)

Q4 All inscribed angles intercepting the same arc are congruent.

Q5 Every angle inscribed in a semicircle is a right angle.

EXPLORE MORE

15. The expression below is equivalent to the length of the arc. The first part of the expression calculates the fractional part of the circle used by the arc. This fraction multiplies the entire circumference of the circle to find the length of the arc.

$$\left(\frac{\text{arc angle a1}}{360} \right) \cdot (\text{Circumference} \odot AB)$$

16. Construct $\overline{AB}$ and midpoint *C*. Construct circle *CB*. Construct $\overline{AD}$ and $\overline{BD}$, where point *D* is on the circle. Triangle *ADB* is a right triangle because *D* is inscribed in a semicircle.

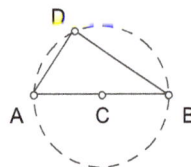

Arcs and Angles

For GSP5

An angle with its vertex at the center of a circle is called a *central angle*. An angle whose sides are chords of a circle and whose vertex is on the circle is called an *inscribed angle*. In this activity you'll investigate relationships among central angles, inscribed angles, and the arcs they intercept.

SKETCH AND INVESTIGATE

1. Construct circle AB.

2. Construct $\overline{AB}$ and make this segment dashed.

Select the segment and choose **Display | Line Style | Dashed.**

3. Construct $\overline{AC}$, where point C is a point on the circle.

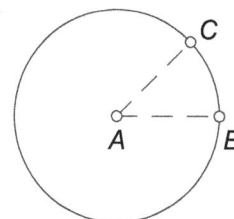

You've just created central angle BAC. Points B and C divide the circle into two arcs. The shorter arc is called a *minor arc* and the longer one is called a *major arc*. A minor arc is named after its endpoints. In the figure above right, the central angle BAC intercepts $\overset{\frown}{BC}$, where $\overset{\frown}{BC}$ is the minor arc.

Select, in order, point B, point C, and the circle. Then, in the Construct menu, choose **Arc on Circle.**

4. Construct the arc on the circle from point B to point C. While the arc is selected, make it thick.

5. Drag point C around the circle to see how it controls the arc. When you're finished experimenting, locate point C so that the thick arc is a minor arc.

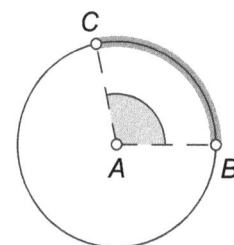

Select $\overset{\frown}{BC}$; then, in the Measure menu, choose **Arc Angle.**

6. Measure the arc angle of $\overset{\frown}{BC}$.

7. Measure $\angle BAC$.

Drag the **Marker** tool from center point A into the angle. Then select the angle marker and choose **Measure | Angle.**

8. Drag point C around the circle again and observe the measures. Pay attention to the differences when the arc is a minor arc and when it is a major arc.

Q1 Write a conjecture about the measure of the central angle and the measure of the minor arc it intercepts.

Q2 Write a conjecture about the measure of the central angle and the measure of the major arc.

9. Construct $\overline{DC}$ and $\overline{DB}$, where point D is a point on the circle, to create inscribed angle CDB.

10. Measure $\angle CDB$.

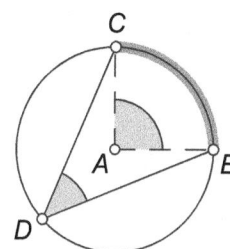

11. Drag point *C* and observe the measures of the arc angle and ∠*CDB*.

Q3 Write a conjecture about the measures of an inscribed angle and the arc it intercepts.

12. Drag point *D* (but not past point *C* or point *B*) and observe the measure of ∠*CDB*.

Q4 Write a conjecture about all the inscribed angles that intercept the same arc.

13. Drag point *C* so that the thick arc is as close to being a semicircle as you can make it.

14. Drag point *D* and observe the measure of ∠*CDB*.

Q5 Write a conjecture about angles inscribed in a semicircle.

EXPLORE MORE

15. In a new sketch, construct a circle and an arc on the circle. Measure the circumference of the circle, the arc angle, and the arc length. Use the circumference and arc angle measurements to calculate an expression equal to the arc length. Explain what you did.

16. Use your conjecture in Q5 to come up with a method for constructing a right triangle. Describe your method.

New York City Title I High School Activities with The Geometer's Sketchpad
© 2012 Key Curriculum Press

Properties of Reflection

ACTIVITY NOTES

SKETCH AND INVESTIGATE

Q1 Point C' traces the mirror image of the student's name.

Q2 Reflection preserves lengths and angle measures.

Q3 A figure and its reflected image are always congruent.

Q4 The vertices of the triangle CDE go from C to D to E in a counterclockwise direction. The vertices of the reflected triangle $C'D'E'$ go from C' to D' to E' in a clockwise direction.

Q5 The mirror line is the perpendicular bisector of any segment connecting a point and its reflected image.

EXPLORE MORE

17. Here is one way to perform this construction: Construct a line through the given point, perpendicular to the given mirror line. Then construct a circle from the intersection of the line and the perpendicular to the original point. The other intersection of the circle and the perpendicular is the reflected image of the original point. For an extra challenge, try doing this construction using only the Euclidean tools, that is, not using the menu at all.

18. Reflect a point across a line. Connect the point with its image point. Also connect each of these points with a third point on the line.

19. Students will notice that they have to flip one of the triangles in order to match the other exactly. Only one triangle will have the shaded side facing up. This is another demonstration that the triangles have opposite orientations.

20. This construction creates an isosceles trapezoid. If the segment connecting two points on one side of the mirror is parallel to the mirror, the points and their images form a rectangle. If the distance from each point to the mirror line is half the distance between the points, the points and their images form a square.

21. Note that this construction creates an angle that is defined by its bisector. It doesn't so much involve bisecting an angle as it does doubling one.

22. The ruler and its image should form a straight line. The apparent distance from your nose to your nose's mirror image is twice the length of the ruler.

Properties of Reflection

When you look at yourself in a mirror, how far away does your image in the mirror appear to be? Why is it that your reflection looks just like you, but backward? Reflections in geometry have some of the same properties of reflections you observe in a mirror. In this activity you'll investigate the properties of reflections that make a reflection the "mirror image" of the original.

SKETCH AND INVESTIGATE: MIRROR WRITING

1. Construct vertical line *AB*.

2. Construct point *C* to the right of the line.

Double-click the line.

3. Mark $\overrightarrow{AB}$ as a mirror.

4. Reflect point *C* to construct point *C'*.

*Select the two points; then, in the Display menu, choose **Trace Points**. A check mark indicates that the command is turned on. Choose **Erase Traces** when you wish to erase your traces.*

5. Turn on **Trace Points** for points *C* and *C'*.

6. Drag point *C* so that it traces out your name.

Q1 What does point *C'* trace?

7. For a real challenge, try dragging point *C'* so that point *C* traces out your name.

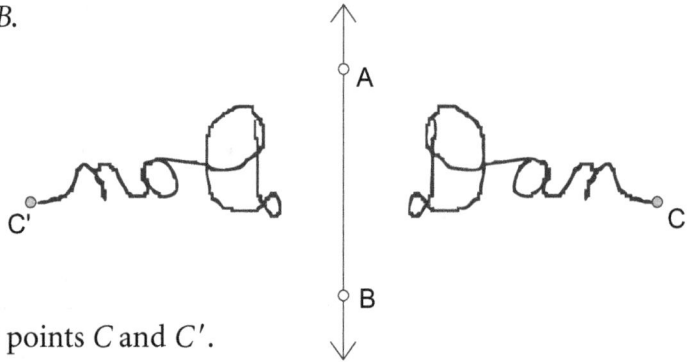

SKETCH AND INVESTIGATE: REFLECTING GEOMETRIC FIGURES

*Select points *C* and *C'*. In the Display menu, you'll see **Trace Points** checked. Choose it to uncheck it.*

8. Turn off **Trace Points** for points *C* and *C'*.

9. In the Display menu, choose **Erase Traces**.

*Select the entire figure; then, in the Transform menu, choose **Reflect**.*

10. Construct △*CDE*.

11. Reflect △*CDE* (sides and vertices) over $\overrightarrow{AB}$.

12. Drag different parts of either triangle and observe how the triangles are related. Also drag the mirror line.

13. Measure the lengths of the sides of triangles *CDE* and *C'D'E'*.

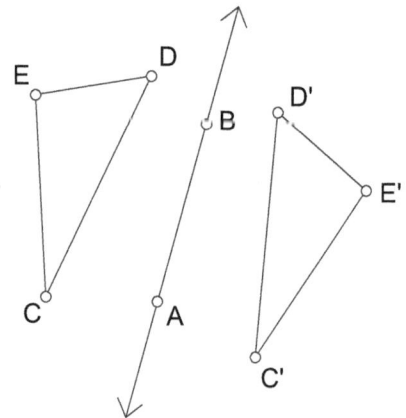

Select three points that name the angle, with the vertex your middle selection. Then, in the Measure menu, choose **Angle.**

14. Measure one angle in $\triangle CDE$ and measure the corresponding angle in $\triangle C'D'E'$.

Q2 What effect does reflection have on lengths and angle measures?

Q3 Are a figure and its mirror image always congruent? State your answer as a conjecture.

Your answer to Q4 demonstrates that a reflection reverses the *orientation* of a figure.

Q4 Going alphabetically from C to D to E in $\triangle CDE$, are the vertices oriented in a clockwise or counterclockwise direction? In what direction (clockwise or counterclockwise) are vertices C', D', and E' oriented in the reflected triangle?

Line Style is in the Display menu.

15. Construct segments connecting each point and its image: C to C', D to D', and E to E'. Make these segments dashed.

You may wish to construct points of intersection and measure distances to look for relationships between the mirror line and the dashed segments.

16. Drag different parts of the sketch around and observe relationships between the dashed segments and the mirror line.

Q5 How is the mirror line related to a segment connecting a point and its reflected image?

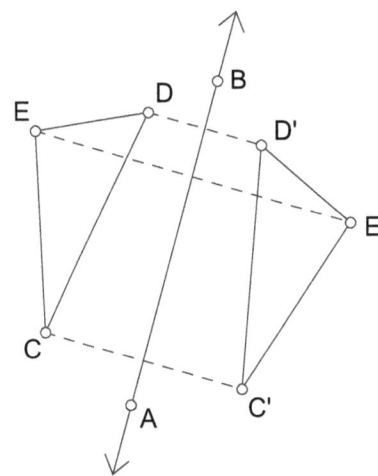

EXPLORE MORE

17. Suppose Sketchpad didn't have a Transform menu. How could you construct a given point's mirror image over a given line? Try it. Start with a point and a line. Come up with a construction for the reflection of the point over the line using just the tools and the Construct menu. Describe your method.

18. Use a reflection to construct an isosceles triangle. Explain what you did.

19. Construct the interiors of the triangles in your sketch and print the sketch. Cut out your triangles and place them both in front of you, shaded side up. Pick up one triangle and place it on top of the other to show that they're congruent. What do you have to do with the triangle you pick up? Do they both still have their shaded sides facing up?

20. Reflect two points across a line and construct the quadilateral with the four points (the two points and their images) as vertices. What kind of quadrilateral is this? How would you have to locate the two points to create a rectangle? A square? Can you create a parallelogram that is not a rectangle? If not, why not?

21. Construct a ray to mark as a reflection mirror. Construct another ray so that when you reflect it across the first, the reflection axis serves as an angle bisector.

22. Look in a mirror. Stand so that your nose is one ruler's length away from the mirror. How can you use the reflection of the ruler to tell whether you're measuring the shortest distance from your nose to the mirror? How far away does your reflected nose appear to be from your actual nose?

Reflections in the Coordinate Plane

SKETCH AND INVESTIGATE

Q1 The y-coordinates of the point and its image are the same. The x-coordinates are opposites. A point with coordinates (a, b) has image coordinates $(-a, b)$. (Make sure students drag an original vertex into different quadrants so they can see these coordinates take on negative values.)

Q2 This time, the x-coordinates of the point and its image are the same and the y-coordinates are opposites. A point with coordinates (a, b) has image coordinates $(a, -b)$.

EXPLORE MORE

1. The coordinates of the image of a point after a reflection across the line $y = x$ are reversed. So a point with coordinates (a, b) has image coordinates (b, a).

2. Both coordinates change sign. A point with coordinates (a, b) has second-image coordinates $(-a, -b)$. The second image is a rotation of 180° about the origin.

Reflections in the Coordinate Plane

In this activity you'll investigate what happens to the coordinates of points when you reflect them across the *x*- and *y*-axes in the coordinate plane.

In the Graph menu, first choose **Show Grid,** then choose **Snap Points.**

1. Show the grid and turn on point snapping.

2. Draw △*CDE* with vertices on the grid.

3. Measure the coordinates of each vertex.

Double-click the axis to mark it as a mirror.

4. Mark the *y*-axis as a mirror.

Select the entire figure; then, in the Transform menu, choose **Reflect.**

5. Reflect the triangle.

6. Measure the coordinates of the image's vertices.

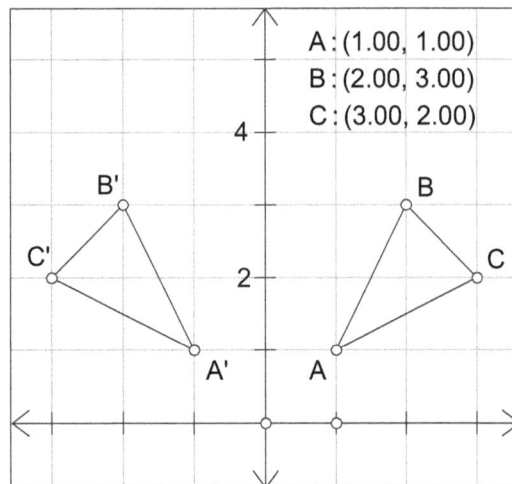

A : (1.00, 1.00)
B : (2.00, 3.00)
C : (3.00, 2.00)

7. Drag vertices to different points on the grid and look for a relationship between a point's coordinates and the coordinates of the reflected image across the *y*-axis.

Q1 Describe any relationship you observe between the coordinates of the vertices of your original triangle and the coordinates of their reflected images across the *y*-axis.

8. Now mark the *x*-axis as a mirror and reflect your original triangle.

9. Before you measure coordinates, can you guess what they'll be? Measure to confirm.

Q2 Describe any relationship you observe between the coordinates of the original points and the coordinates of their reflected images across the *x*-axis.

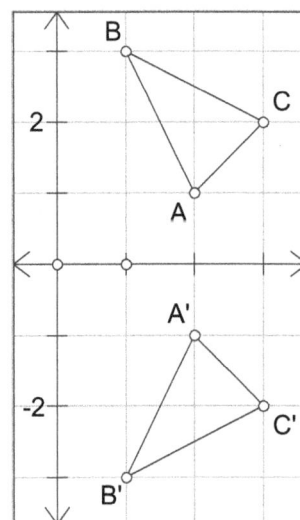

EXPLORE MORE

1. Draw a line on the grid that passes through the origin and makes a 45° angle with the *x*-axis (in other words, the line *y* = *x*). Reflect your triangle across this line. What do you notice about the coordinates of the vertices of this image?

2. Experiment with a reflection across the y-axis followed by a reflection of the images across the x-axis. What can you say about the coordinates of this product of two reflections? How is this second image related to the original triangle?

Translations in the Coordinate Plane

SKETCH AND INVESTIGATE

Q1 If you drag point B along the x-axis, the y-coordinates of a vertex and its image point will be equal, but the x-coordinates will differ.

Q2 If you drag point B along the y-axis, the x-coordinates of a vertex and its image point will be equal, but the y-coordinates will differ.

Q3 If the vector translates the triangle up and to the left, point B is in the second quadrant. Its x-coordinate is negative (causing the movement to the left) and its y-coordinate is positive (causing the movement up).

Q4 The image of (x, y) under a translation by (a, b) is the point $(x + a, y + b)$.

EXPLORE MORE

1. These two translations combined are equivalent to a single translation by a vector from the origin to the second vector tip.

Translations in the Coordinate Plane

In this activity you'll investigate what happens to the coordinates of points when they're translated in the coordinate plane.

In the Graph menu, first choose **Show Grid**; then choose **Snap Points**.

1. Show the grid and turn on point snapping.

Using the **Text** tool, click on a point to display its label. Double-click the label to change it.

2. Draw a segment from the origin to anywhere on the grid. Label the origin *A* and the other endpoint *B*.

Select point *A* and point *B*, in that order; then, in the Transform menu, choose **Mark Vector**. Watch for the animation indicating the marked vector.

3. Measure the coordinates of point *B*.

4. Mark vector *AB*.

5. Draw $\triangle CDE$ with vertices on the grid.

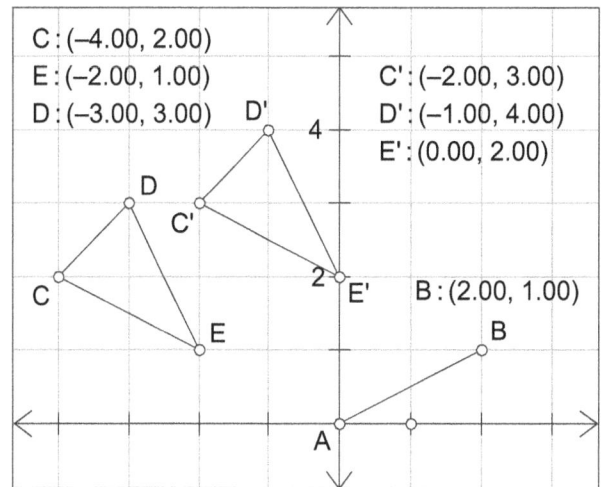

C : (–4.00, 2.00)
E : (–2.00, 1.00)
D : (–3.00, 3.00)
C' : (–2.00, 3.00)
D' : (–1.00, 4.00)
E' : (0.00, 2.00)
B : (2.00, 1.00)

Select the entire figure; then, in the Transform menu, choose **Translate**.

6. Translate the triangle by the marked vector.

7. Measure the coordinates of the two triangles' six vertices.

8. Experiment by dragging point *B* or any of the triangle vertices. Look for a relationship between a point's coordinates and the coordinates of its image under a translation.

Q1 Where can you drag point *B* so that the original points and the corresponding image points always have the same *y*-coordinates but have different *x*-coordinates?

Q2 Where can you drag point *B* so that the original points and the corresponding image points always have the same *x*-coordinates but have different *y*-coordinates?

Q3 When the vector defined by the origin and point *B* translates your original triangle to the left and up, what must be true of the coordinates of point *B*?

Q4 Suppose point *B* has coordinates (*a*, *b*). What are the coordinates of the image of a point (*x*, *y*) under a translation by (*a*, *b*)?

EXPLORE MORE

1. In the same sketch, draw a segment from point *C* to a new point on the grid. Mark a new vector from point *C* to this point, and then translate $\triangle D'E'F'$ by this new vector. You'll get a triangle $D''E''F''$ that is the result of two translations from the original triangle *DEF*. Find a vector that will translate $\triangle DEF$ to $\triangle D''E''F''$ in a single translation. How is this single translation vector related to the first two?

Reflections across Two Parallel Lines

ACTIVITY NOTES

SKETCH AND INVESTIGATE

Q1 Two reflections across a pair of parallel lines are equivalent to a translation.

Q2 $AA'' = 2EF$

Q3 $\overline{EF}$ is perpendicular to both lines. Because parallel lines are everywhere equidistant, EF represents the distance between them.

Q4 The translation along the vector AA' moves the original figure onto the second reflected image. This demonstrates that reflecting a figure across two parallel lines is equivalent to a single translation. The translation vector is perpendicular to the lines and is twice the distance between the lines.

Q5 a. $AE = EA'$

   b. $A'F = FA''$

   c. $AA' + A'A'' = AA''$

   d. Because $AE = EA'$, $(1/2)A'A = EA'$.
   Because $A'F = FA''$, $(1/2)A'A'' = A'F$.
   Also, the distance between the parallel lines, $EF = EA' + A'F$.
   Substituting from above gives:

$$EF = (1/2)A'A + (1/2)A'A'' = (1/2)AA''$$

EXPLORE MORE

1. If you reverse the order of the lines in the two reflections, the translation goes in the opposite direction.

2. Construct a segment connecting a point with its translated image. Construct the perpendicular bisector of this segment and a line parallel to the perpendicular bisector through one of the segment's endpoints. Reflections across these two parallel lines will be equivalent to the translation.

New York City Title I High School Activities with The Geometer's Sketchpad
© 2012 Key Curriculum Press

213

Reflections across Two Parallel Lines

For GSP5

In this investigation you'll see what happens when you reflect a figure across a line then reflect the image across a second line parallel to the first.

SKETCH AND INVESTIGATE

Hold down the Shift key while you click with the **Point** tool so that points will stay selected as you construct them. Then, in the Construct menu, choose **Quadrilateral Interior.**

With the **Text** tool, click on a point to display its label. Double-click a label to change it.

Double-click the line to mark it as a mirror. Select the polygon interior and the point; then, in the Transform menu, choose **Reflect.**

Select point *D* and $\overrightarrow{BC}$; then, in the Construct menu, choose **Parallel Line.**

1. Construct any irregular polygon interior.

2. Show the label of one of the polygon's vertices. Change the label to *A* (if necessary).

3. Construct a line *BC*.

4. Mark the line as a mirror and reflect the polygon and point *A* across it.

5. Construct point *D* and a line through point *D* parallel to $\overleftrightarrow{BC}$.

6. Mark this second line as a mirror and reflect the first reflected image and point *A'* across it.

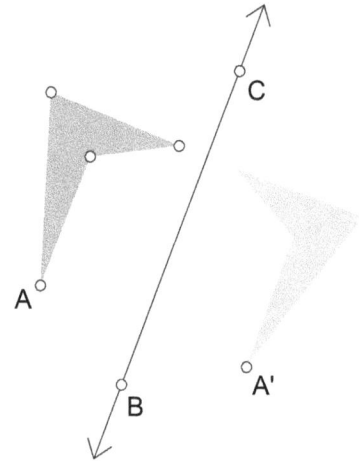

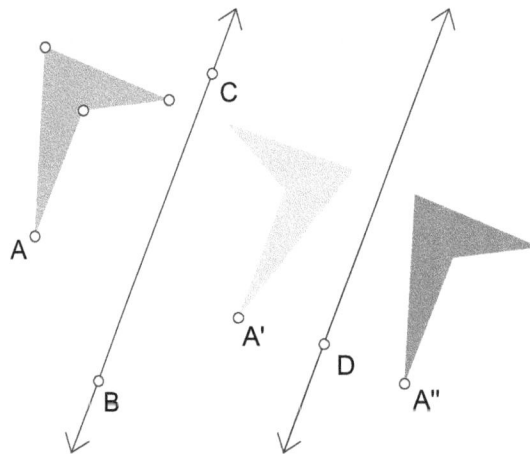

7. Drag the original figure and the two lines and observe their relationships to the two images.

Q1 Two reflections move your original figure to its second image. What single transformation do you think will do the same thing? (If you're not sure, go on to the next steps, then come back to this question.)

8. Construct $\overline{AA''}$.

9. Construct points E and F where $\overline{AA''}$ intersects the two lines.

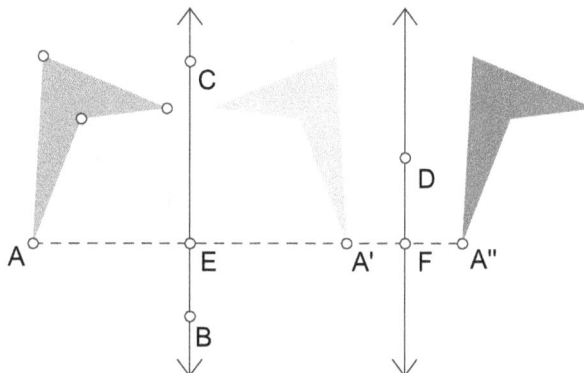

Select points A and A''; then, in the Measure menu, choose **Distance.** Repeat for EF.

10. Measure AA'' and EF.

Q2 Drag one of the lines and compare the two distances. How are they related?

Q3 EF is the distance between the two lines. Why?

Select points A and A'' in order; then, in the Transform menu, choose **Mark Vector.** Select the figure; then, in the Transform menu, choose **Translate.**

11. Mark AA'' as a vector, then translate the original figure by this vector.

Q4 Describe the result of step 11, above. What single transformation is equivalent to the combination of two reflections across parallel lines?

Q5 Answer the following questions to explain why AA'' and EF are related as they are:

a. How does AE compare to EA'? _____

b. How does $A'F$ compare to FA''? _____

c. $AA' + A'A'' =$ _____

d. Complete the rest of the explanation on your own.

EXPLORE MORE

1. In the same sketch, try reflecting your figure and then its image across the two lines in the opposite order. Describe the result.

2. In a new sketch, construct a figure and translate it by some arbitrary distance. Construct two lines so that you can move your original figure onto its translated image in two reflections.

Reflections across Two Intersecting Lines   For GSP5 ACTIVITY NOTES

SKETCH AND INVESTIGATE

Q1 The combination of two reflections across intersecting lines is equivalent to a single rotation.

Q2 $m\angle ABA'' = 2 \cdot m\angle CBD$.

Q3 A rotation about the point of intersection of two lines is equivalent to a combination of two reflections across the lines. The angle of rotation is twice one of the angles formed by the intersecting lines.

Q4 a. $m\angle ABC = \frac{1}{2} m\angle ABA'$

b. $m\angle A'BD = \frac{1}{2} m\angle A'BA''$

c. $m\angle ABA' + m\angle A'BA'' = m\angle ABA''$

d. Since $m\angle ABC = m\angle CBA'$, and $m\angle CBA' + m\angle A'BD = m\angle CBD$, then, by substitution, $m\angle ABC + m\angle A'BD = m\angle CBD$. $\frac{1}{2} m\angle ABA' + \frac{1}{2} m\angle A'BA'' = m\angle CBD$, therefore, $\frac{1}{2} m\angle ABA'' = m\angle CBD$.

EXPLORE MORE

12. Changing the order of the two reflections reverses the direction of the rotation (from clockwise to counterclockwise or vice versa).

13. The image after being reflected across a pair of parallel lines is a translation.

14. Construct a segment connecting a point and its rotated image. Construct the perpendicular bisector. Repeat for another pair of points. The center of rotation is located where the two perpendicular bisectors intersect.

Reflections across Two Intersecting Lines

In this investigation you'll see what happens when you reflect a figure across a line and then reflect the image across a second line that intersects the first.

SKETCH AND INVESTIGATE

Use the **Polygon Interior** tool. Finish the polygon by clicking the first point again.

1. Construct any irregular polygon.

2. Show the label of one of the polygon's vertices and change it to *A* (if necessary).

3. Construct two intersecting lines and their point of intersection.

With the **Text** tool, click a point to display its label. Double-click a label to change it.

Double-click the line to mark it as a mirror. Select the polygon and the point; then, in the Transform menu, choose **Reflect**.

4. Mark the line closest to the polygon as a mirror; then reflect the polygon and the labeled point across this line. Change the color of the reflected image. (See the figure below. If necessary, move the polygon so that the image falls between the lines.)

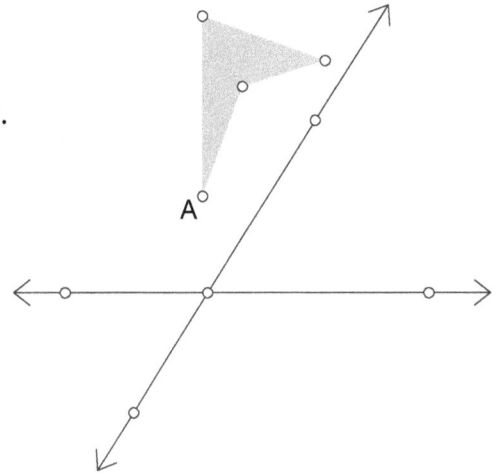

Steps 1–3

5. Mark the other line as a mirror, and then reflect the image from the first reflection across this second line. Change the color of this second image. (See the figure below.)

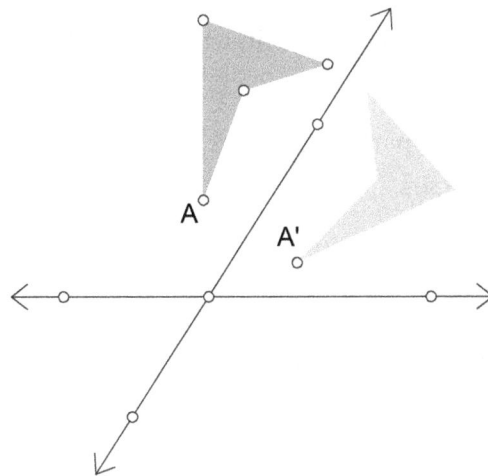

Step 4

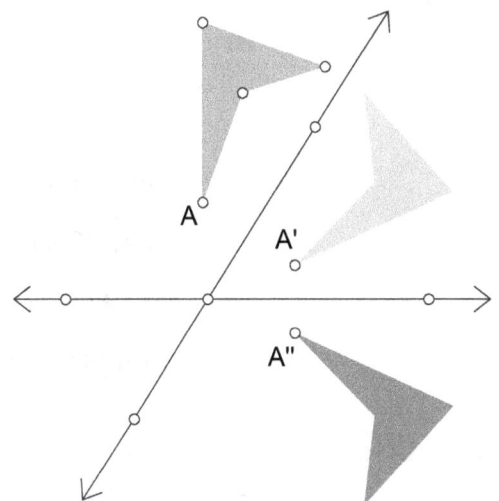

Step 5

6. Drag the original figure and the two lines and observe their relationships to the two images.

Q1 Two reflections move your original figure to its second image. What single transformation do you think will do the same thing? (If you're not sure, go on to the next steps, and then come back to this question.)

7. Construct $\overline{AB}$, where point A is a point on the original figure and point B is the point of intersection of the lines.

8. Construct $\overline{BA'}$ and $\overline{BA''}$.

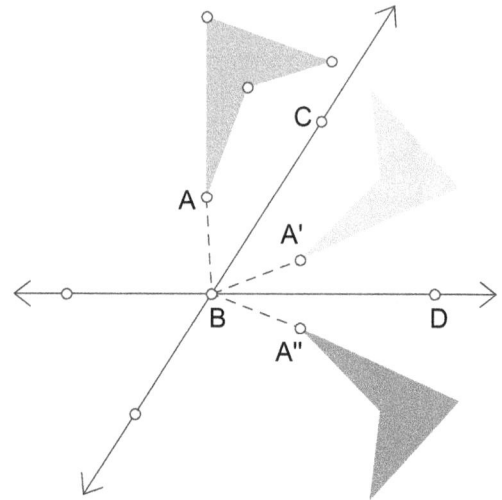

Select the three points that name the angle, with the vertex your middle selection; then, in the Measure menu, choose **Angle.**

9. Measure $\angle ABA''$.

10. Measure $\angle CBD$, the angle between the lines.

Q2 Compare the two angle measures. How are they related?

Double-click point B to mark it as a center. Select, in order, points A, B, and A''; then, in the Transform menu, choose **Mark Angle.** Select the figure; then, in the Transform menu, choose **Rotate.**

11. Mark point B as a center for rotation and mark $\angle ABA''$; then rotate the original figure by this angle.

Q3 Describe the result of step 11. What single transformation is equivalent to the combination of two reflections across intersecting lines?

Q4 Answer the following questions to explain why m$\angle ABA''$ and m$\angle CBD$ are related the way they are:

    a. How does m$\angle ABC$ compare to m$\angle ABA'$? _____ (Try to answer without measuring first.)

    b. How does m$\angle A'BD$ compare to m$\angle A'BA''$? _____

    c. m$\angle ABA'$ + m$\angle A'BA''$ = m$\angle$ _____

    d. Complete the rest of the explanation on your own.

EXPLORE MORE

12. Continuing in the same sketch, try reflecting your figure and then its image across the two lines in the opposite order. Describe the result.

13. In the same sketch, drag the points on the lines so that the lines don't intersect any more. (The angle measures and the segments from the point of intersection should disappear.) Now how is the original figure related to the second image? What's the result of combining reflections across two parallel lines?

14. In a new sketch, construct a figure and a center of rotation. Rotate the figure about the center by any fixed angle. Hide the center and see if you can locate two intersecting lines so that reflections across them move your original figure to the rotated image.

SKETCH AND INVESTIGATE

Q1 A glide reflection is the product of a reflection and a translation.

Q2 A translation is the product of two reflections across parallel lines.

Q3 A glide reflection is the product of three reflections. Two of the reflection lines can be parallel, although they do not have to be (see Explore More 3, below).

EXPLORE MORE

10. First, split point *G* from line *EF*. (Select point *G* and choose **Split Point from Line** from the Edit menu.) Then select, in order, the original polygon, point *E*, point *F*, point *G*, and the first glide-reflected image. Choose **Create New Tool** from the Custom Tools menu (the bottom icon in the Toolbox). Name the tool "Glide Reflection." Now merge point *G* back to line *EF*. To use the new tool, click the **Custom** tool icon in the Toolbox, and then click the objects in the sketch as instructed in the Status Line. To make the tool easier and quicker to use, choose **Show Script View** from the Custom Tools menu, double-click point *E* in the Given list, and check **Automatically match sketch object** in the dialog box that appears. Do the same for points *F* and *G*. From now on, students will only have to click the pre-image to use the tool (as long as the mirror line is defined by points labeled *E* and *F* and there's a point on the line labeled *G*).

11. This takes some time, but has fun results and is a good introduction to action buttons. Students can complete it as a long-term assignment or project. The Presentation can be refined by choosing the individual Hide/Show buttons, choosing **Properties** from the Edit menu, and changing settings on the Hide/Show panel.

12. If students try reflecting across three random lines, they will discover that this is a glide reflection as well (though it may not be obvious). They can confirm this by selecting the image, the original, and the three lines and defining the three reflections as a custom tool. When they apply this repeatedly, they'll see the glide reflection pattern.

13. A rotation followed by a reflection is a glide reflection (or, in the special case where the center of rotation is a point on the reflection line, a simple reflection). A rotation followed by a translation is a rotation by a different angle with a different center.

Glide Reflections

In this activity you will investigate an isometry called a *glide reflection*. A glide reflection is not a transformation found in the Transform menu, but you'll define it as a custom tool, and in the process you'll learn what a glide reflection is and what it does.

SKETCH AND INVESTIGATE

Hold down the Shift key while you click with the **Point** tool so that points will stay selected as you construct them. Then, in the Construct menu, choose **Quadrilateral Interior.**

Double-click the line to mark it as a mirror. Select the polygon interior; then, in the Transform menu, choose **Reflect.**

Select, in order, points *E* and *G;* then, in the Transform menu, choose **Mark Vector.** A brief animation indicates the mark. Select the reflected polygon interior; then, in the Transform menu, choose **Translate.**

Line *EF* and vector *EG* will remain marked; you don't have to mark them over and over again.

1. Construct an irregular polygon interior, like polygon *ABCD*, shown at right.

2. Construct $\overleftrightarrow{EF}$.

3. Mark $\overleftrightarrow{EF}$ as a mirror and reflect the polygon interior across $\overleftrightarrow{EF}$.

4. Construct a point *G* on the line so that *E* and *G* are about an inch apart.

5. Mark $\overrightarrow{EG}$ as a vector and translate the reflected image by the marked vector. This second image is a glide reflection of your original figure.

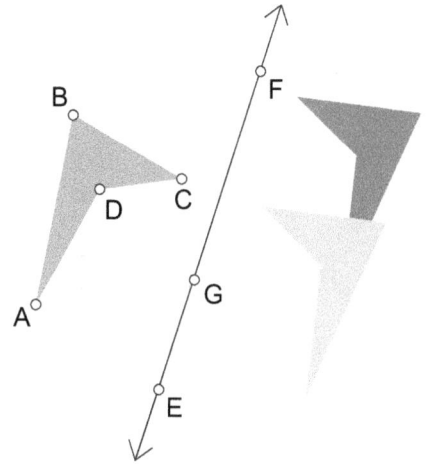

6. Drag point *G* to see how it affects the glide-reflected image.

7. Hide the intermediate image (the first reflection).

8. Use the techniques you've learned so far to create two more glide-reflected images, as shown at right.

9. Drag parts of your sketch (vertices of the original polygon, point *G*, points *E* and *F*, the line) and observe the effects.

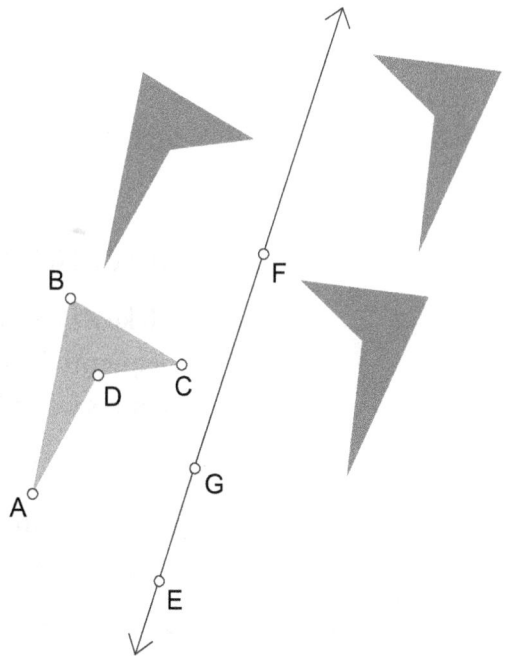

Q1 A glide reflection is the product of two transformations. What are they?

Q2 A translation is also the product of two transformations. What are they?

Q3 A glide reflection can be thought of as a product of what three transformations?

EXPLORE MORE

For tips on making and using custom tools, choose **Tools** from the Help menu, and then click the Custom Tools link.

10. Turn your glide reflection into a custom tool to save yourself time when doing glide reflections. Then save the sketch (which includes your new tool) in your own personal tool folder. Check with your teacher if you have questions about how to create your personal tool folder.

11. Create a polygon that looks like a foot; then use glide reflections to make a sketch that looks like footprints in sand. To make the footprints appear sequentially, as if made by a walking invisible person, follow these steps.

    a. Make a Hide/Show action button for each footprint.

    b. Make a Presentation button with the Hide/Show buttons. Choose **Sequentially** in the Presentation dialog box that appears, and add a one-second Pause Between Actions.

    c. Hide all of the footprints, and then press the *Presentation* button.

12. Experiment with reflections across three random lines. Does this produce a glide reflection?

13. What's the product of a rotation and a reflection? What's the product of a rotation and a translation?

SKETCH AND INVESTIGATE

Q1 Every line of symmetry passes through the center of the polygon. If the regular polygon has an odd number of sides, each line of symmetry is a perpendicular bisector of one side and passes through an opposite vertex. If the regular polygon has an even number of sides, each line of symmetry either is a perpendicular bisector of a pair of opposite sides or passes through a pair of opposite vertices.

Q2 Students will fill in one entry in row 2 here, depending on which polygon they were working with.

# of sides of regular polygon	3	4	5	6
# of reflection symmetries	3	4	5	6
# of rotation symmetries	3	4	5	6

Q3 Answers will vary depending on the student's polygon. Also, for any given regular polygon, there are many correct answers. To check a student's answer, calculate $360°/n$, where n is the number of sides in the polygon, and make sure their rotation angle is some integer multiple of this quotient.

Q4 Students will fill in one entry in row 3 in the chart shown in Q2, depending on which polygon they were working with.

Q5 Students complete the chart shown in Q2.

Q6 A regular n-gon has n axes of reflection symmetry and n rotation symmetries. The smallest angle of rotation symmetry larger than 0° is $360°/n$.

EXPLORE MORE

16. Answers will vary. A quick way to construct a regular polygon is to rotate by a central angle. For example, to construct a pentagon, construct two points and rotate one point around the other five times, by 72° each time.

17. Have students pick a shape, define the shape, and report on its symmetry. Make sure they address special cases, such as rectangles that are also squares, because these cases can have more symmetry. This Explore More can make a nice long-term assignment or project. Students might describe their results as a poster, or create a sketch with color, text, a Show/Hide button, and animation.

Symmetry in Regular Polygons

A figure has *reflection symmetry* if you can reflect the figure across a line so that the image will coincide with the original figure. The line you reflect across is called a *line of symmetry* or a *mirror line*. A figure has *rotational symmetry* if you can rotate it some number of degrees about some point so that the rotated image will coincide with the original figure. In this exploration you'll look for reflection and rotation symmetries of regular polygons.

SKETCH AND INVESTIGATE

Construct your polygon from scratch or use a custom tool. The sketch **Polygons.gsp** includes tools for regular polygons.

1. Construct a regular polygon and its interior. You can use an equilateral triangle, a square, a regular pentagon, or a regular hexagon. You may want to have different groups in your class investigate different shapes.

2. Construct a line.

Double-click the line to mark it as a mirror. Select the polygon interior; then, in the Transform menu, choose **Reflect**.

3. Mark the line as a mirror and reflect the polygon interior over it.

4. Give the image a different color.

5. Drag the line until the image of your polygon coincides exactly with the original.

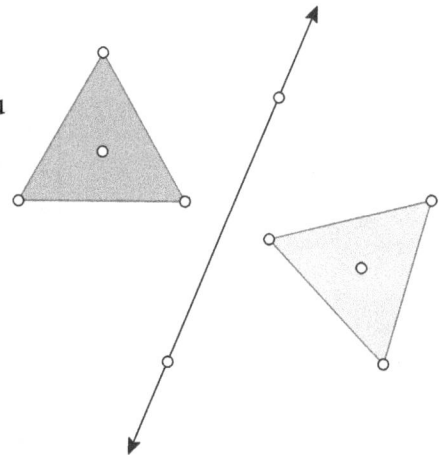

Q1 When a reflection image coincides with the original figure, the reflection line is a line of symmetry. Describe how the line of symmetry is positioned relative to the figure.

6. Drag the line so that it is a different line of symmetry. Repeat until you have found all the reflection symmetries of your polygon.

Q2 Fill in one entry in the table here: the number of reflection symmetries for your polygon. (*Note:* Be careful not to count the same line twice!) You'll come back to fill in other entries as you gather more information.

Number of sides of regular polygon	3	4	5	6	. . .	n
Number of reflection symmetries					. . .	
Number of rotation symmetries					. . .	

Next, you'll look for rotation symmetries.

Select the three points, with the vertex your middle selection. Then, in the Measure menu, choose **Angle.**

7. Move the line so that the reflected image is out of the way.

8. If the polygon's center doesn't already exist, construct it.

$m\angle ABC = 30.00°$

9. Use the **Segment** tool to construct an angle.

10. Measure the angle.

Choose **Preferences** from the Edit menu and go to the Units panel.

Double-click the point to mark it. Select the angle measurement; then, in the Transform menu, choose **Mark Angle.** Select the interior; then, in the Transform menu, choose **Rotate.**

11. In Preferences, set Angle Units to **directed degrees**.

12. Mark the center of the polygon as a center for rotation and mark the angle measurement. Rotate the polygon interior by this marked angle measurement.

13. Give the rotated image a different color.

14. Change the angle so that the rotated image fits exactly over the original figure.

Q3 What angle measure causes the figures to coincide?

Polygon: _____   Rotation angle: _____

15. Continue changing your angle to find all possible rotation symmetries of your polygon.

Q4 Count the number of times the rotated image coincides with the original when rotating from 0° to 180° and from −180° back to 0°. In your chart from Q2, record the total number of rotation symmetries you found. (*Note:* Count your initial position, or 0°, as one of your rotation symmetries.)

Q5 Combine the results from other members of your class to complete your chart with the reflection and rotation symmetries of other regular polygons.

Q6 Use your findings to write a conjecture about the reflection and rotation symmetries of a regular *n*-gon. Include in your conjecture a statement about the smallest angle of rotational symmetry greater than 0.

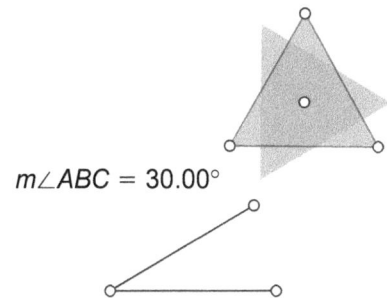

EXPLORE MORE

16. Construct one of the regular polygons using rotations or some combination of rotations and reflections. Explain your method and how it takes advantage of the symmetry of the polygon.

17. Explore symmetries in other figures. For example, try rhombuses, rectangles, isosceles trapezoids, or kites. Report your findings.

9

Exponents, Radicals, and Complex Numbers (Algebra II and Trigonometry)

Zero and Negative Exponents

POSITIVE EXPONENTS

2. The tools will work for any nonzero setting of a, but the activity creates an exponential sequence, so the heights of the rectangles can get out of hand if a is much more than 1.5.

Q1 $a^2 \cdot a = a^3$

Q2 Multiplying by a raises the power by one: $a^n \cdot a = a^{n+1}$. This is true because $a^n = a \cdot a \cdot a \cdots a$, where a appears as a factor n times. If you multiply this by one more a, there will be $n + 1$ factors.

Q3 As the exponents increase, the heights of the bars do not increase by the same amount each time. Provided $a > 1$, they increase by more and more each time. This is more obvious if you change the scale by dragging one of the tick numbers on the axis. Students may use specific examples to explain: "Multiplying 1 by 1.37 adds only 0.37 to the bar height, but multiplying 10 by 1.37 results in 13.7, adding 3.7 to the original height."

ZERO EXPONENT

Q4 $a^4 \div a = a^3$, and generally, $a^n \div a = a^{n-1}$. As in Q2, $a^n = a \cdot a \cdot a \cdots a$, where a appears as a factor n times. If you divide this by a, you cancel the last factor, leaving $(n - 1)$ factors.

Q5 $a^0 = 1$. This follows from the fact that $a^1 \div a = 1$.

NEGATIVE EXPONENTS

Q6 $a^0 \div a = a^{-1}$, using the rule from Q3.

Q7 Since $1 = a^0$, dividing by a three times is equivalent to dropping the exponent three times to a^{-3}. Therefore, $1/a^3 = a^{-3}$.

EXPLORE DIFFERENT BASES

In this extension students can use the same sketch to see the effects of changing the base.

Q8 If $a > 1$, higher exponents always correspond to larger rectangles, hence, larger numbers. As you keep multiplying, there is no limit to how high the bars will go. Similarly, if you divide repeatedly, there is no limit to how short the bars will become.

Q9 If $0 < a < 1$, higher exponents correspond to smaller numbers. This follows from the fact that multiplying any positive number by a number greater than one increases it, while multiplying it by a number between one and zero makes it smaller. In either case, the result is always positive.

Q10 As a changes, any bar with the value a^0 remains constant. This is because $a^0 = 1$.

Q11 When a is less than zero, the bars alternate between positive and negative. The numbers with odd exponents are negative. Those with even exponents are positive.

In each step of the activity, students formed the next number by either multiplying or dividing by a negative number, a. Multiplying or dividing by a negative changes the sign, thus creating the alternating pattern.

Q12 Problems and solutions to the Simplify game vary.

(Both the activity document and the presentation document have another custom tool, **Measure Bar,** which was not used in the activity. Choose the tool and click one of the bars. It will display the number represented by the bar.)

Zero and Negative Exponents

In this presentation students observe the visual pattern formed when an exponent increases as a number is repeatedly raised to higher powers, and observe the related pattern as the exponent is reduced first to zero and then to negative values.

Explain that the label on the bar is a^1, which is the same thing as a.

1. Open **Zero Exponents Present.gsp.** Drag the marker so students can see how it changes the value of a. Return the marker to its original position.

2. To multiply the value represented by the first bar by a, press and hold the **Custom** tool icon to display the Custom Tools menu. Choose the **Multiply By a** tool. This tool works by itself, so there is no need to click anything.

   > Create New Tool...
   > Tool Options...
   > Show Script View
   > -- This Document --
   > Multiply By a
   > Divide By a
   > Measure Bar

Q1 Ask students what the new bar represents. Drag the marker to change the value of a to 2, so that students can see the new bar has the value 4. Return the marker to its original position before continuing.

Q2 Ask, "What will be the result if we multiply a^2 by a?"

3. Choose **Multiply By a** again. Use the tool several times.

Q3 Ask, "As the exponents increase, do the heights of the bars increase by the same amount each time? How can you tell?"

4. Choose the custom tool **Divide By a.**

Q4 Ask students to explain why the new bar is the height that it is.

5. Use **Divide By a** several more times, until the progression runs down to a^0.

Q5 Ask, "What do you think is the value of a^0?" Drag the marker so that students can see that the value of a has no effect on this bar.

Q6 Ask, "What will happen if we divide by a again?"

6. Choose **Divide By a.**

Q7 Discuss what the resulting bar represents. Choose **Divide By a** twice more during the discussion. Try to get students to propose formulations like these:

$$a^{-1} = 1 \div a = \frac{1}{a} \quad \text{and} \quad a^{-3} = 1 \div a \div a \div a = 1 \cdot \frac{1}{a} \cdot \frac{1}{a} \cdot \frac{1}{a} = \frac{1}{a^3}$$

Use the other numbered pages to create different patterns that involve both positive and negative exponents.

Use the "Simplify" page to give students practice in manipulating expressions to eliminate negative exponents.

Zero and Negative Exponents

By now you should be comfortable doing calculations with exponents that are positive integers. From here, certain questions naturally arise. What if the exponent is zero? What if it is negative? What if it is not an integer? This activity explores the concept of zero and negative exponents. Non-integer exponents will have to wait.

POSITIVE EXPONENTS

1. Open **Zero Exponents.gsp.**

The bar represents the number a. You can drag the marker to change its value. The label on the bar is a^1, which is the same thing as a.

2. Start with a between 1 and 1.5. You can change it later. Now multiply a^1 by a. Press and hold the **Custom** tool icon to display the Custom Tools menu. Choose the **Multiply By a** tool. This tool works by itself, so there is no need to click anything.

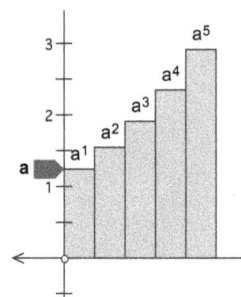

Another vertical bar appears representing a^2. You get this result because $a^1 \cdot a = a^2$.

Q1 What is the result if you multiply a^2 by a?

3. Choose **Multiply By a** again. Use the tool several times and study the result.

Q2 Consider the number a^n, where n is a positive integer. What happens when you multiply the number by a? State a general rule. Explain why this is true.

Q3 As the exponents increase, do the heights of the bars increase by the same amount each time? How can you tell? Explain your observations.

ZERO EXPONENT

4. Go to page 2. This is the same sketch. The progression of bars goes up to a^4.

5. Choose the custom tool **Divide By a.** It does just what the name says.

Q4 What is $a^4 \div a$? What happens when you divide a^n by a? Explain why this is true.

6. Use the **Divide By a** tool two more times, so that the progression runs down to a^1. Dividing by a once more should give you a^0. Try it.

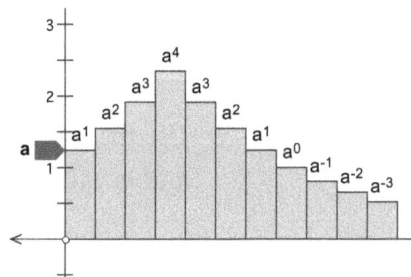

Q5 What is the value of a^0? Drag the marker to test different values of a.

NEGATIVE EXPONENTS

Q6 At this point, a^0 should be the last number in the progression you are building. Using your answer to Q3, what will be the result when you divide by a again? Choose **Divide By a** and check your answer.

Q7 Starting with the number 1, if you divide by a three times, that is the same as dividing by a^3.

$$1 \div a \div a \div a = 1 \cdot \frac{1}{a} \cdot \frac{1}{a} \cdot \frac{1}{a} = \frac{1}{a^3}$$

How can you write this same number with a negative exponent? Use the sketch to check your answer.

EXPLORE DIFFERENT BASES

The point of this activity was to investigate zero and negative exponents, but you may have noticed some interesting changes that occur when you change the base, a. Drag the marker to change the value of a, and answer the following questions.

Q8 Start with $a > 1$. If you keep multiplying the bar lengths by a, is there a limit to how high the bars will go? If you keep dividing them by a, is there a limit to how short the bars will become?

Q9 Pull the marker downward so that $0 < a < 1$. What change do you see in the pattern formed by the bars? Explain why this is.

Q10 As you change a, which bars do not change at all? Why?

Q11 When a is negative, you will see an entirely different pattern. Describe the pattern, and explain why it looks this way.

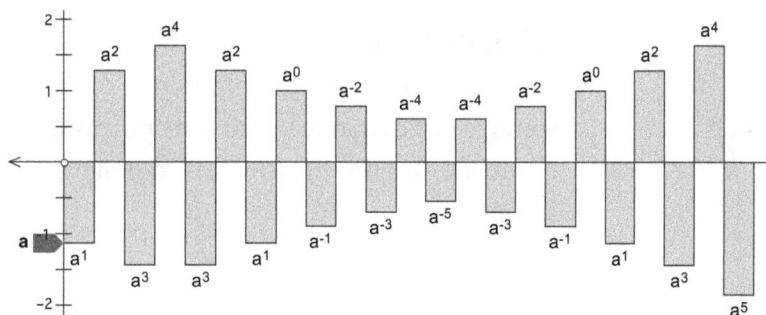

Q12 Play the game on the "Simplify" page at least four times. Each time, write down the original problem and the solution. Try writing the solution down before you actually drag the variables.

The Square Root Spiral

SKETCH AND INVESTIGATE

Q1 The Pythagorean theorem gives you $1^2 + 1^2 = c^2$, which simplifies to $c^2 = 2$, or $c = \sqrt{2}$.

Q2 Triangle $A''A'B$ is a right triangle, so you can use the Pythagorean theorem to calculate the length of its hypotenuse. Leg $A'B$ has length 1 (it's the radius of a circle with radius 1) and leg $A'A''$ has length $\sqrt{2}$. So we have $1^2 + (\sqrt{2})^2 = c^2$, or $3 = c^2$, or $c = \sqrt{3}$.

Q3 and **Q4**

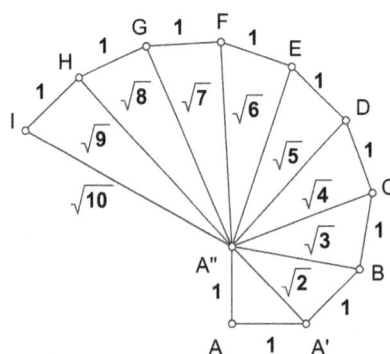

Q5

$\sqrt{2} \approx 1.414$	$\sqrt{3} \approx 1.732$	$\sqrt{4} = 2$
$\sqrt{5} \approx 2.236$	$\sqrt{6} \approx 2.449$	$\sqrt{7} \approx 2.646$
$\sqrt{8} \approx 2.828$	$\sqrt{9} = 3$	$\sqrt{10} \approx 3.162$

EXPLORE MORE

1. When students have made their custom tools, this construction takes very little extra time and provides a dynamic spiral. (The original spiral is not dynamic because its purpose is to model specific irrational numbers.) It's fun for students to make action buttons to animate their spirals.

The Square Root Spiral

Irrational numbers such as $\sqrt{2}$ and $\sqrt{3}$ correspond to points on a ruler, but you can't find those points precisely by dividing your ruler into fractional parts. However, you can construct square roots with compass and straightedge (or with Sketchpad). In this activity you'll construct a square root spiral and use it to create a chart of approximate square roots.

SKETCH AND INVESTIGATE

The first part of the spiral is an isosceles right triangle.

Choose **Preferences** from the Edit menu and go to the Units panel.

1. In Preferences, set the Distance Units to **inches** and set Distance Precision to **thousandths**.

With point *A* selected, choose **Translate** in the Transform menu.

2. Construct point A and translate it at an angle of 0° by 1 inch to create point A'.

Double-click point *A* to mark it as a center. Select point *A'*; then, in the Transform menu, choose **Rotate**.

3. Mark point A as a center and rotate point A' by 90°.

4. Connect these points to make isosceles right triangle $AA'A''$.

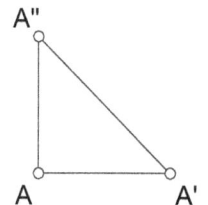

Q1 Use the Pythagorean theorem to find the exact value of $A'A''$. (Use radical form, not a decimal approximation, and show your work.)

In steps 5–9, you'll construct $\sqrt{3}$.

5. Construct a line through point A', perpendicular to $\overline{A'A''}$.

6. Construct circle $A'A$.

7. Construct point B, the intersection of the circle with the line, as shown.

8. Hide the circle and line.

9. Construct $\overline{A'B}$ and $\overline{A''B}$.

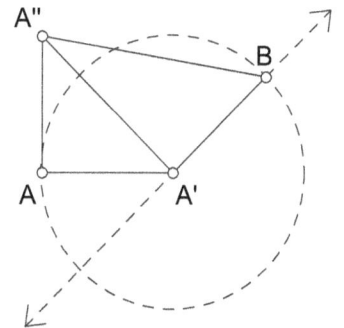

Q2 Explain why $\overline{A''B}$ is equal to $\sqrt{3}$ inches.

You can continue in this way to construct a spiral that gives you the square roots of as many consecutive positive integers as you like. A custom tool makes the construction process quicker. Follow steps 10–17 to continue your spiral.

10. Construct a line through point *B*, perpendicular to $\overline{BA''}$.

11. Construct circle $\overline{BA'}$.

12. Construct point *C*, the intersection of the circle with the line, as shown.

13. Hide the circle and line.

14. Construct segments $A''C$ and CB.

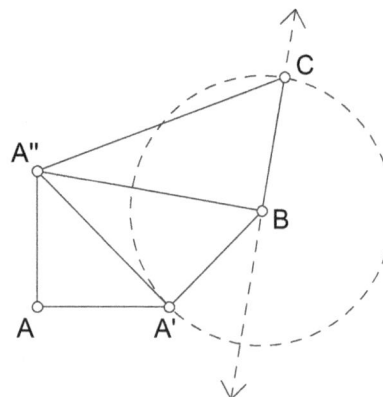

Selection order is very important to make the custom tool work.

15. Now create a custom tool. Select, in order, point *A′*, point *B*, point *A″* (these are the tool's givens), point *C*, $\overline{A''C}$, and $\overline{BC}$ (these are the tool's *results*). Choose **Create New Tool** from the Custom Tools menu (the bottom tool in the Toolbox) and name the tool **Next Triangle.**

16. Click on the **Custom** tools icon to select the tool you just created. Click, in order, on point *B*, point *C*, and point *A″* to create the next right triangle.

17. Use your custom tool to create at least five more right triangles.

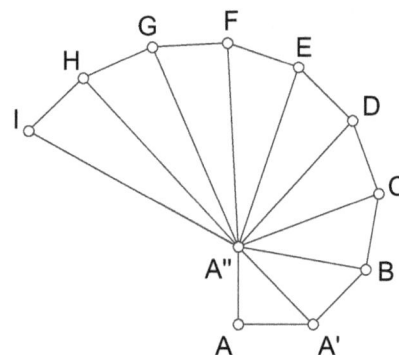

Save this sketch. You can use its custom tool again in Explore More 1, below.

Q3 What are the lengths of all the segments around the outside of the spiral? Write these lengths on the figure at right.

Q4 Using radical form, write the lengths of the spiral arms on the figure above right.

Q5 Measure all the spiral arms, then complete the table below with approximations to the nearest thousandth for square roots of whole numbers from 2 to 10.

$\sqrt{1} = 1$	$\sqrt{2} \approx$	$\sqrt{3} \approx$	$\sqrt{4} =$	$\sqrt{5} \approx$
$\sqrt{6} \approx$	$\sqrt{7} \approx$	$\sqrt{8} \approx$	$\sqrt{9} =$	$\sqrt{10} \approx$

EXPLORE MORE

1. Draw an arbitrary triangle. Use the **Next Triangle** tool on its three vertices. Now use the tool, in the same order, on the vertices of the newly created triangle. Do this at least eight more times. You should get a spiral (though not a square root spiral) that you can open and close dynamically by dragging the vertices of your original triangle.

Multiplication of Complex Numbers

For GSP5 ACTIVITY NOTES

SKETCH AND INVESTIGATE

Q1 The vector $\overrightarrow{OF}$ represents **wz.**

Q2 The length of $\overrightarrow{OC}$ is $a|\mathbf{z}|$.

Q3 The length of $\overrightarrow{OD}$ is $|\mathbf{z}|$, and the length of $\overrightarrow{OE}$ is $b|\mathbf{z}|$.

Q4 The vector $\overrightarrow{CF}$ is a translation of $\overrightarrow{OE}$, so its length is the same: $b|\mathbf{z}|$.

Q5 Vector $\overrightarrow{OE}$ was obtained by multiplying $\overrightarrow{OZ}$ by i and then dilating it by b. Multiplication by i corresponds to a rotation of 90°. Thus $\overrightarrow{CF}$, a translation of $\overrightarrow{OE}$, makes a 90° angle with $\overrightarrow{OC}$.

Q6 Both $\triangle OVW$ and $\triangle OCF$ are right triangles. The sides OC and CF of $\triangle OCF$ are each $|\mathbf{z}|$ times as long as the sides OV and VW of $\triangle OVW$. Thus, by the SAS triangle similarity theorem, the two triangles are similar.

Q7 The scale factor of the two similar triangles is $|\mathbf{z}|$. Thus the length of $\overrightarrow{OF}$ is $|\mathbf{w}| \cdot |\mathbf{z}|$.

Q8 Because $\triangle OVW$ and $\triangle OCF$ are similar, both $\angle WOV$ and $\angle COF$ equal α. Thus the argument of $\overrightarrow{OF}$ is $\alpha + \beta$.

Q9 To multiply two complex numbers w and z, *multiply* their lengths and *add* their arguments.

Q10 The length of **wz** is 6, and its argument is 150°. Drawing a picture and applying the relationships found in a 30-60-90 triangle yields $\mathbf{wz} = -3\sqrt{3} + 3i$.

EXPLORE MORE

Q11 The complex number with length 1 and argument 270° equals -1 when squared. This number is $-i$.

Q12 The complex number with length 1 and argument 45° will equal i when squared. Written in the form $x + iy$, this number is $\frac{\sqrt{2}}{2} + \frac{i\sqrt{2}}{2}$. Similarly, a complex number with length 1 and argument 225° will also equal i when squared. This number is $\frac{-\sqrt{2}}{2} - \frac{i\sqrt{2}}{2}$.

Q13 The complex numbers $a + bi$ and $a - bi$ are reflections of each other across the real axis. Their arguments sum to 0° (equivalently, 360°) when the numbers are multiplied. Thus the product sits on the positive real axis.

Multiplication of Complex Numbers

By using the distributive property of multiplication and the fact that $i^2 = -1$, you can use algebra to compute the product of two complex numbers, $a + bi$ and $c + di$:

$$(a + bi)(c + di) = ac + (bc + ad)i + bdi^2$$

$$= (ac - bd) + (bc + ad)i$$

Algebra alone doesn't tell the whole story. In this activity you'll explore complex number multiplication from a geometric perspective.

SKETCH AND INVESTIGATE

1. Open **Complex Multiplication.gsp.** You'll see the complex number $\mathbf{w} = a + bi$ and another complex number, $\mathbf{z}$ on the complex plane. Because complex numbers have both a length and a direction with respect to the origin, $\mathbf{w}$ and $\mathbf{z}$ are represented as vectors. Your goal is to construct the product, $\mathbf{zw}$.

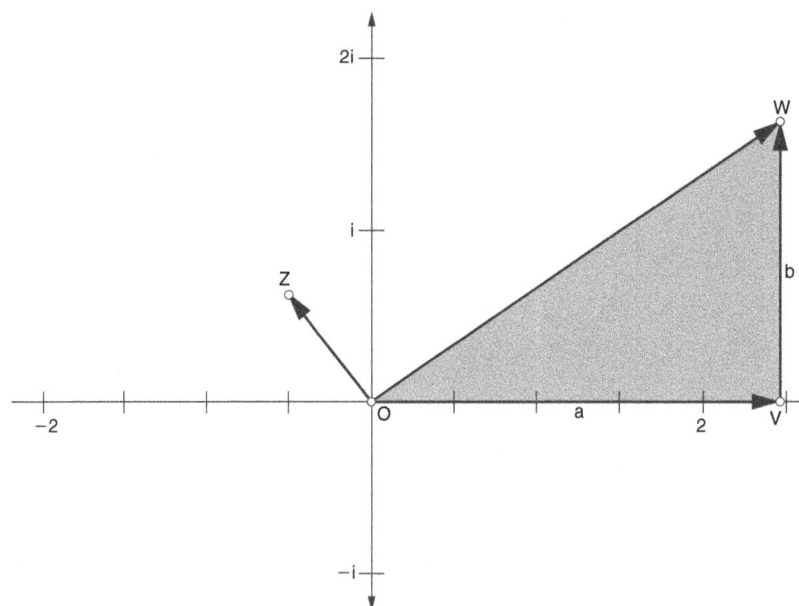

Algebraically, $\mathbf{zw} = \mathbf{z}(a + bi) = \mathbf{z}a + \mathbf{z}bi$. In finer detail, multiplying $\mathbf{w}$ by $\mathbf{z}$ consists of four tasks:

    Task 1: Multiply vector $\mathbf{z}$ by the scalar value a.
    Task 2: Multiply vector $\mathbf{z}$ by i.
    Task 3: Multiply the resulting vector $\mathbf{z}i$ by the scalar value b.
    Task 4: Add vector $\mathbf{z}a$ and vector $\mathbf{z}bi$.

The sketch comes with three pre-built tools for carrying out these steps visually on the complex plane. The tools, available by clicking the **Custom** tool icon in the Toolbox, are listed here:

- **Scale:** Multiplies a vector (complex number) by a scalar value. To use the tool, click the head of a vector and then the desired scalar value.

- **Multiply by i:** Multiplies a vector (complex number) by i. To use the tool, click the head of the vector.

- **Add complex #s:** Adds together two vectors (complex numbers) and displays the result as a vector. To use the tool, click the heads of the vectors.

2. Use the custom tools to multiply **w** by **z**. Follow, in order, the four tasks listed earlier. Your completed construction should look similar to the following picture:

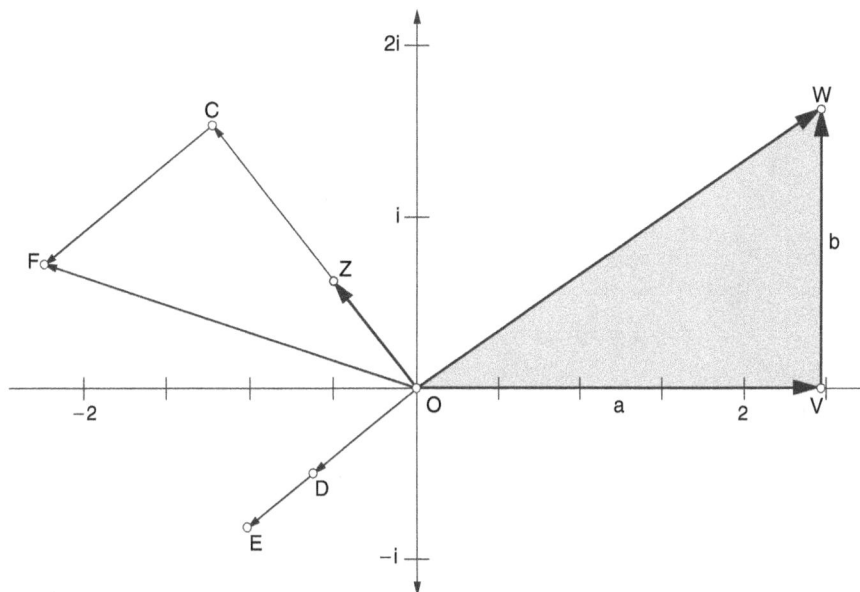

Q1 Which vector represents **wz** in the picture?

Triangle OVW has sides of lengths a, b, and $|w|$, and OZ has length $|z|$. Your answers to the next three questions should all be given in terms of these values.

Q2 What is the length of $\overrightarrow{OC}$?

Q3 What are the lengths of $\overrightarrow{OD}$ and $\overrightarrow{OE}$?

Q4 What is the length of $\overrightarrow{CF}$?

Q5 What is the measurement of $\angle OCF$?

Q6 Explain why $\triangle OVW$ is similar to $\triangle OCF$.

Remember, multiplication by i is equivalent to a 90° rotation.

Q7 Given that the two triangles are similar, what is the length of $\overrightarrow{OF}$?

Look back at your
similar triangles.

Q8 The argument of a vector is the angle it makes with the positive real axis. If the argument of $\overrightarrow{OW}$ is α and the argument of $\overrightarrow{OZ}$ is β, then what is the argument of $\overrightarrow{OF}$?

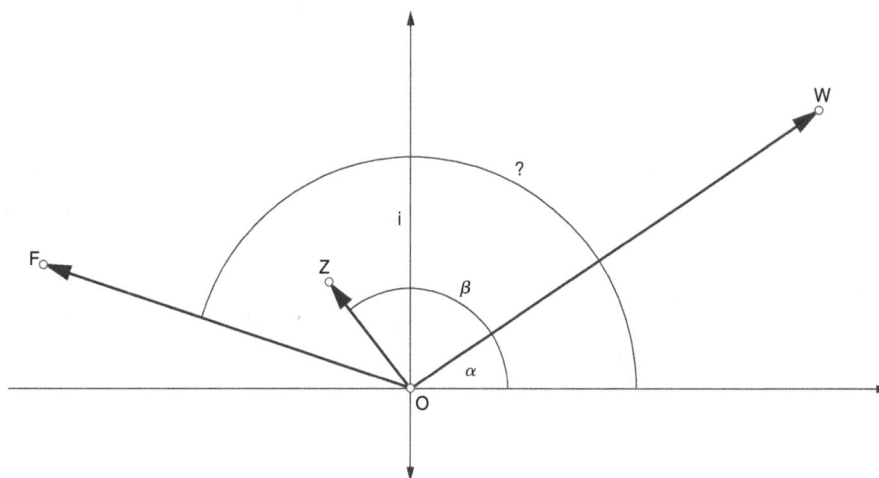

Q9 Complete this sentence:

To multiply two complex numbers **w** and **z**, _____ their lengths and _____ their arguments.

Q10 If $|\mathbf{w}| = 3$, $|\mathbf{z}| = 2$, $\alpha = 40°$, and $\beta = 110°$, then find the value of **wz**. Write your answer in the form $x + iy$.

EXPLORE MORE

Now that your sketch is complete, you can use it to explore the product of any two complex numbers.

3. Hide all vectors in your picture, leaving only $\overrightarrow{OZ}$, $\overrightarrow{OW}$, and $\overrightarrow{OF}$.

Q11 Find a complex number other than i whose square is equal to -1.

Q12 Find two different complex numbers a and b such that $a^2 = b^2 = i$.

Q13 Explain geometrically why the product of a complex number $a + bi$ and its reciprocal $a - bi$ is a real number.

In Search of Buried Treasure

ACTIVITY NOTES

SKETCH AND INVESTIGATE

Before turning to Sketchpad, you can ask students to model the treasure directions on your classroom floor or in the schoolyard. Mark the location of the two trees and pick a random spot for the scarecrow. Use string or pace out steps to find and mark the treasure. Do so again for a new location of the scarecrow. Students should discover that the treasure is in roughly the same location in both cases.

Q1 The treasure, point *T*, stays in the same spot regardless of the location of point *S*. Surprisingly, then, it doesn't matter where the scarecrow once stood. The walking directions lead to the same spot for every possible scarecrow location you might choose.

PROVE YOUR RESULTS

Q2 Point *T* is at *i* on the complex plane. Two other noteworthy observations are

- The origin is equidistant from the oak tree, the elm tree, and the treasure.

- The treasure is on the perpendicular bisector of the segment connecting the oak and elm trees.

Q3 Moving the origin 1 unit to the left shifts the real component of a complex number by $+1$. Thus the new location of the scarecrow is $(a + 1) + bi$.

Q4 Multiplication by *i* corresponds to a rotation of 90°.

Q5 The location of M_1 is *iS*, which equals $-b + i(a + 1)$.

Q6 Moving the origin 1 unit back to the right shifts the real component of a complex number by -1. Thus the new location of M_1 is $-b - 1 + i(a + 1)$.

Q7 The new location of the scarecrow is $(a - 1) + bi$.

Q8 Multiplication by $-i$ corresponds to a rotation of $-90°$.

Q9 The location of M_2 is $-iS$, which equals $b - i(a - 1)$.

Q10 The new location of M_2 is $b + 1 - i(a - 1)$.

New York City Title I High School Activities with The Geometer's Sketchpad
© 2012 Key Curriculum Press

241

Q11 To find the midpoint of M_1M_2, add M_1 to M_2 and then compute the average of the real and imaginary components. $M_1 + M_2 = [-b - 1 + i(a + 1)] + [b + 1 - i(a - 1)] = 0 + 2i$. Dividing by 2 yields i. Thus, as seen in the sketch, the treasure sits at i.

PROJECT IDEAS

Other proofs of the treasure problem also exist. Ask students to research the proofs found in the February 1998 and September 2003 issues of *Mathematics Teacher* magazine.

In Search of Buried Treasure

Tucked away in the attic of your grandparents' farmhouse is a dusty old letter describing the location of a buried treasure. It reads:

> To find the treasure, walk into the field behind the farmhouse. Start at the scarecrow. Walk straight from the scarecrow to the oak tree. Count your steps as you walk. When you reach the oak tree, turn 90 degrees to the right and count off the same number of steps. When you reach the end, place a marker in the ground.
>
> Now return to the scarecrow. Walk straight from the scarecrow to the elm tree. Again, count your steps as you walk. When you reach the elm tree, turn 90 degrees to the left and count off the same number of steps. When you reach the end, place a marker in the ground.
>
> Connect the two markers with a rope. Dig beneath the midpoint of the rope to find the treasure.

The oak tree and the elm tree are still in the field, but the scarecrow is long gone. Is the treasure lost?

In this activity you'll unearth the buried treasure by building a Sketchpad model and then prove your results by using complex numbers.

SKETCH AND INVESTIGATE

1. Open a new sketch and draw three random points to represent the scarecrow (point S), the oak tree (point O), and the elm tree (point E).

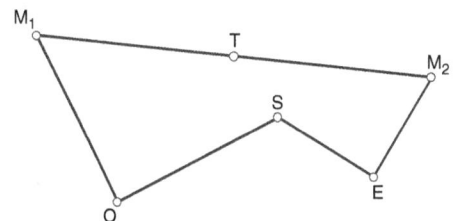

2. Draw segments OS and SE.

3. Select point O and choose **Transform | Mark Center.**

Choose **Rotate** from the Transform menu.

4. Select segment OS (including point S) and rotate it 90° about point O. Label the marker (the rotated point S) M_1.

To add a subscript at the end of a label, type the subscript enclosed in square brackets—for instance, M[1].

5. Rotate segment ES by $-90°$ about point E. Label the second marker M_2.

6. Construct point T—the location of the treasure—the midpoint of segment M_1M_2.

Q1 Drag point S around the screen. What do you notice about the treasure, point T?

PROVE YOUR RESULTS

7. Open **Buried Treasure.gsp.** You'll see an arrangement similar to your own sketch, except in this sketch, the points and segments have been placed on the complex plane.

The axes have been arranged so that the oak tree and the elm tree sit at $(-1, 0)$ and $(1, 0)$, respectively.

Q2 Drag point S and observe whether point T moves. What is the location of point T on the complex plane?

Point S is free to move anywhere on the complex plane. We'll describe its location as (a, b) or, equivalently, $a + bi$.

8. Press the *Move Origin to Oak Tree* button, and watch the origin reposition itself 1 unit to the left.

Q3 Prior to moving the origin, the location of the scarecrow was $a + bi$. What is its new location?

Q4 By construction, M_1 is a rotation of point S by 90° about O. In the complex plane, multiplication by what number corresponds to a rotation of 90°?

Q5 Use your answer to Q4 to write the location of point M_1.

9. Press the *Move Origin Back* button, and watch the origin move 1 unit to the right, back to its original location.

Q6 What is the new location of point M_1?

10. Press the *Move Origin to Elm Tree* button, and watch the origin reposition itself 1 unit to the right.

Q7 Prior to moving the origin, the location of the scarecrow was $a + bi$. What is its new location?

Q8 By construction, point M_2 is a rotation of point S by $-90°$ about point E. In the complex plane, multiplication by what number corresponds to a rotation of $-90°$?

Q9 Based on your answer to Q8, what is the location of point M_2?

11. Press the *Move Origin Back* button, and watch the origin move 1 unit to the left, back to its original location.

Q10 What is the new location of point M_2?

Remember, T is the midpoint of segment M_1M_2.

Q11 Based on your answers to Q6 and Q10, compute the location of the treasure. Does your result match what you see in the sketch?

Powers of Complex Numbers

PRELIMINARY WORK

Q1 To multiply two complex numbers, multiply their lengths and add their arguments.

Q2 To square a complex number, square its length and double its argument.

Q3 $2i = 2(\cos 90 + i\sin 90)$

Q4 Based on Q2 and Q3, $z = \sqrt{2}\,(\cos 45 + i\sin 45)$ is equal to $2i$ when squared.

INVESTIGATE z^2

Q5 $z = 1 + i$

To position z as precisely as possible (by a tenth of a pixel at a time), students can use the movement buttons.

Note: For most of the remaining Sketchpad-related questions, even moving by a tenth of a pixel at a time does not allow students to obtain an exact answer. You can ask students how to calculate precise answers after they obtain their Sketchpad approximations. (Trigonometry is of great help.)

Q6 The complex number $-(1 + i)$ also equals $2i$ when squared since $[-(1 + i)]^2 = (-1)^2 \cdot (1 + i)^2 = 1 \cdot 2i$.

If students get stuck on this question, encourage them to drag z around while looking for different positions that place z^2 at $2i$.

Q7 $z = \sqrt{2}\,(\cos 225 + i\sin 225)$

Q8 Squaring the value of z from Q7 yields $z^2 = 2(\cos 450 + i\sin 450) = 2i$.

HIGHER POWERS OF z

Q9 Written in the form $a + bi$, the three values of z are $\sqrt[3]{2}\left(\sqrt{3}/2 + 1/2i\right)$, $\sqrt[3]{2}\left(-\sqrt{3}/2 + 1/2i\right)$, and $-\sqrt[3]{2}\,i$. Sketchpad will express these values in decimal notation.

Written in the form $r(\cos\theta + i\sin\theta)$, the three values of z are $\sqrt[3]{2}(\cos 30 + i\sin 30)$, $\sqrt[3]{2}(\cos 150 + i\sin 150)$, and $\sqrt[3]{2}(\cos 270 + i\sin 270)$.

Q10 To cube a complex number z, cube its length and triple its argument.

Q11 Cubing $\sqrt[3]{2}$ and tripling the angle measures of the values from Q9 yields three complex numbers that equal $2i$.

Q12 The four values of z are $\sqrt[4]{2}(\cos 22.5 + i\sin 22.5)$, $\sqrt[4]{2}(\cos 112.5 + i\sin 112.5)$, $\sqrt[4]{2}(\cos 202.5 + i\sin 202.5)$, and $\sqrt[4]{2}(\cos 292.5 + i\sin 292.5)$.

Q13 The five values of z are $\sqrt[5]{2}(\cos 18 + i\sin 18)$, $\sqrt[5]{2}(\cos 90 + i\sin 90)$, $\sqrt[5]{2}(\cos 162 + i\sin 162)$, $\sqrt[5]{2}(\cos 234 + i\sin 234)$, and $\sqrt[5]{2}(\cos 306 + i\sin 306)$.

Q14 On page 1, the two red x's form a segment passing through the origin. On page 2, three red x's form the vertices of an equilateral triangle. On page 3, the x's form a square, and on page 4, the x's form a regular pentagon.

EXPLORE MORE

Q15 In addition to using Sketchpad, students can apply their results from previous questions to solve $z^n = -1$. The argument of one z-value satisfying the equality is $180/n$. You can find the remaining $n - 1$ arguments by adding $360/n$ degrees repeatedly to $180/n$. This pattern forms the basis of De Moivre's Theorem.

Powers of Complex Numbers

If $z^2 = 1$, it's not hard to name two values of z that satisfy the equality. Both $z = 1$ and $z = -1$ work.

Now consider $z^3 = -1$. One solution is $z = -1$. There are no other real solutions. Are there other solutions in the complex numbers?

In this activity you'll find solutions to equations like $z^3 = -1$ by exploring the geometric effects of raising complex numbers to integer powers.

PRELIMINARY WORK

There are two common ways to represent a complex number z. The first is $z = a + bi$, where a and b are real numbers.

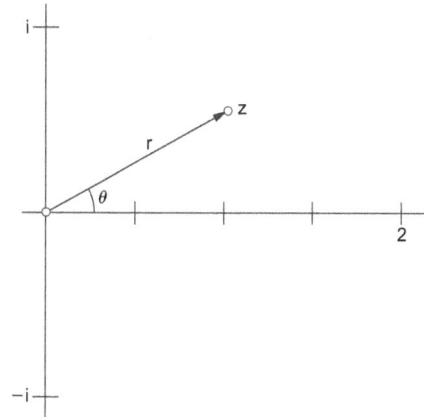

This second form is often written using the shorthand notation $re^{i\theta}$.

Another form is $z = r(\cos\theta + i\sin\theta)$. In this representation, z is a vector with length r that makes an angle of θ with the positive real axis. The angle θ is called the *argument* of z.

Q1 Complete this statement:

To multiply two complex numbers, _____ their lengths and _____ their arguments.

Q2 Complete this statement based on Q1:

To square a complex number z, _____ its length and _____ its argument.

Q3 Write $2i$ in the form $r(\cos\theta + i\sin\theta)$.

Q4 Use the information from Q2 and Q3 to find a complex number z that equals $2i$ when squared.

INVESTIGATE z^2

1. Open **Complex Powers.gsp.** The first page of the sketch shows the complex plane with real numbers on the horizontal axis and imaginary numbers on the vertical axis. The sketch contains two complex numbers, z and z^2.

Position point z as best you can so that $z^2 = 2i$. Use the arrow keys for small adjustments. Use the movement buttons for even smaller adjustments.

Q5 Drag point z to determine whether the value of z from Q4 does indeed satisfy $z^2 = 2i$. What is the value of z in the form $a + bi$?

If you need a hint, drag z to a different position for which $z^2 = 2i$.

Q6 Real numbers have both a positive and a negative square root. Complex numbers are neither positive nor negative, but a similar idea applies. Can you name another value of z that satisfies $z^2 = 2i$? Write your answer in the form $a + bi$.

Q7 Use your sketch to determine whether the value of z from Q6 satisfies $z^2 = 2i$. What is the value of z in the form $r(\cos\theta + i\sin\theta)$?

Use the method from Q2.

Q8 Check that the $r(\cos\theta + i\sin\theta)$ value of z from Q7 is indeed equal to $2i$ when squared.

2. Position the two red x's to indicate the two locations of z for which $z^2 = 2i$.

HIGHER POWERS OF z

Open page 2 of the sketch **Complex Powers.gsp.** You'll see three complex numbers represented as vectors: z, z^2 (not labeled), and z^3.

Q9 Drag point z to find three values of z for which $z^3 = 2i$. Write each in the form $a + bi$ and $r(\cos\theta + i\sin\theta)$.

Q10 Complete this statement:

To cube a complex number z, _____ its length and _____ its argument.

Use the method from Q10.

Q11 Look at each of the values from Q9 that you wrote in the form $r(\cos\theta + i\sin\theta)$. Cube each to check that $z^3 = 2i$.

3. Position the three red x's to indicate the three locations of z for which $z^3 = 2i$.

Q12 Open page 3 of the sketch. Find and verify the four values of z for which $z^4 = 2i$. Place red x's at these locations.

Q13 Open page 4 of the sketch. Find and verify the five values of z for which $z^5 = 2i$. Place red x's at these locations.

Q14 On all four pages of your sketch, connect consecutive red x's (the solutions to $z^n = 2i$) with segments, proceeding either clockwise or counterclockwise. Describe the shapes formed by the segments.

EXPLORE MORE

Q15 Use the sketch to solve $z^2 = -1$, $z^3 = -1$, $z^4 = -1$, and $z^5 = -1$.

10

Relations and Functions
(Algebra II and Trigonometry)

Direct Variation

In this activity students move from looking at properties of lines (in particular, slope) to generating linear relationships. Before starting, you might have a discussion about quantities that grow in different ways (proportionally, exponentially, and inversely) in relation to each other—and how each of these would look on a graph. The activity Inverse Variation focuses on quantities that are inversely proportional.

This activity focuses on lines of the form $y = mx$ (or $y = bx$), and another activity (The Slope-Intercept Form of a Line) focuses on lines of the form $y = mx + b$. It would be valuable to make the connection here between direct relationships and linear relationships (of which direct relationships are a subset).

SKETCH AND INVESTIGATE

1. If Sketchpad is set to its default Preference settings, points won't be labeled when they are created. Students can click points with the **Text** tool to label them. (Points will be labelled in alphabetical order.) To edit a label, double-click it with the **Text** tool.

Q1 Dragging B changes only the base and the area. Dragging C changes only the height and the area.

12. The sketch becomes cluttered at this point. Students may want to move the origin down near the bottom of the sketch window, hide the grid, and move the rectangle to a relatively clear area of the sketch.

Q2 It shows that as the height gets bigger, the area gets bigger; that as the height gets smaller, the area gets smaller; and that they grow or shrink proportionally to each other.

Q3 $A = base \cdot height$

Q4 $f(x) = base \cdot x$

Q5 It's the same. (To be more precise, it contains the path of the plotted point; the function exists in the first and fourth quadrants, but the plotted point is always in the first quadrant.)

Q6 The graph passes through the origin because the area of a rectangle with height 0 is 0—hence the point $(0, 0)$. And algebraically, when $x = 0$ in $f(x) = base \cdot x$, $f(0) = 0$—hence the point $(0, 0)$.

Q7 A rectangle can't have a negative height or area. The domain should be restricted to $x > 0$ (or possibly $x \geq 0$ if you consider 0 to be a possible height of a rectangle).

Q8 It means that as one quantity doubles, the other doubles; as one triples or halves, the other triples or halves. For example, the area of a rectangle with base 3 and height 4 is 12. If you double the height to 8 (and leave the base the same), the area also doubles to 24. The word "proportional" is used because the area and the height are in proportion (12/4 = 24/8 = 3). The base is the constant of proportionality.

Q9 This changes the slope of the line. Students may also notice that the plotted point moves vertically up and down (which makes sense because the x-value, which represents height, is not changing).

Q10 The length of the base is the slope of the graph. "Wide" rectangles (those with larger bases) will have steeper graphs. The reason is that every increase in height will add a lot to the area. "Skinny" rectangles (those with smaller bases) will have more gradual graphs. The reason is that similar increases in height will add much less to the area.

EXPLORE MORE

Q11 The point traces out a portion of a parabola. The trace is no longer linear, so this is *not* direct variation. The reason this happens is that we are now varying both the height and the base simultaneously, whereas before we were varying only the height, leaving the base constant. Variation in one dimension results in a linear graph, whereas variation in two dimensions results in a quadratic graph.

WHOLE-CLASS PRESENTATION

Use the sketch **Direct Variation Present.gsp** to explore with the class the relationships between measurements in a rectangle. The goal is to use this visual tool (and hopefully students' intuition) to think about why certain quantities are proportional and to connect this proportionality to direct variation via an equation and a graph.

Use page 1 of the sketch to answer Q1–Q10 together as a class. Use page 2 for Explore More (optional), which features a square instead of a rectangle. In this case, the area also increases when the side length increases, but the relationship is quadratic rather than linear.

Direct Variation

What happens to the area of a rectangle if you keep the length of the base constant while varying the height? (Try to answer this question before reading on.) What happens if you enlarge the entire rectangle? In this activity you will learn about direct variation and how it's represented algebraically and graphically.

SKETCH

Start by constructing a rectangle and its interior.

1. In a new sketch, construct a point and label it *A*.

To translate *A*, select it and choose **Transform | Translate.** To construct the ray, select the two points and choose **Construct | Ray.**

2. Translate point *A* by 1.0 cm at 0°. Construct a horizontal ray from *A* through the translated point.

3. Hide the translated point, construct a new point on the ray, and label it *B*.

4. Translate *B* by 1.0 cm at 90°. Construct a vertical ray from *B* through the translated point.

5. Hide the translated point, construct a new point on the ray, and label it *C*.

Select *B* and *A* in order and choose **Transform | Mark Vector.**

6. Mark the vector from point *B* to point *A*.

7. Translate point *C* by the marked vector. Label the translated point *D*.

8. Hide the rays and construct $\overline{AB}$, $\overline{BC}$, $\overline{CD}$, and $\overline{DA}$.

You should have a rectangle. Drag each of the four points to be sure it remains a rectangle and to see how dragging each point changes the rectangle.

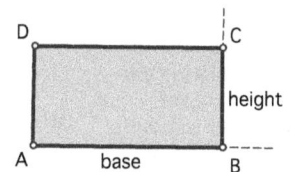

Select the four points in order and choose **Construct | Quadrilateral Interior.**

9. Construct polygon interior *ABCD*. Measure its area.

10. Measure the lengths of base $\overline{AB}$ and height $\overline{BC}$ by selecting them and choosing **Measure | Length.**

11. Click the **Text** tool on $\overline{AB}$ and $\overline{BC}$ to show their labels. Change $\overline{AB}$'s label to *base* and $\overline{BC}$'s label to *height* by double-clicking the **Text** tool on each label.

INVESTIGATE

Next you'll investigate how the area changes as you change the size of the rectangle.

Q1 Drag the points around. How do the measurements change when you drag *B*? How do the measurements change when you drag *C*?

If the sketch is too cluttered, choose **Graph | Hide Grid.**

12. A graph will help to show what's happening. Select in order the *height* and *Area ABCD* measurements, and choose **Graph | Plot as (x, y).**

You have just plotted a point whose *x*- and *y*-coordinates are the height and area of the rectangle, respectively. (If you can't see it yet, you will soon.)

To trace an object, select it and choose **Display | Trace Plotted Point.** To clear traces from the screen, choose **Display | Erase Traces.**

13. Drag point *C* closer to $\overline{AB}$ and observe the effect on the plotted point. Trace the plotted point so you can see what it does as you drag point *C*.

Q2 What does the path of the plotted point tell you about how height and area are related in a rectangle when the base is kept constant?

Q3 Write the formula for the area of a rectangle in terms of *base* and *height*.

Q4 Now write the same formula as a function using *f*(*x*) for *area* and *x* for *height*.

To enter *base* in your formula, click its measurement in the sketch. To enter *x*, click the *x* key on the Calculator's keypad.

Q5 Plot your function from Q4 by choosing **Graph | Plot New Function.** How does the function plot relate to the path of the plotted point as you vary the height of the rectangle?

Q6 Why does it make sense that the graph passes through the origin?

Q7 The function is plotted as a line. But the traces cover only part of the line. Why? If you wanted the function plot to accurately represent the situation, what part of it would you cut off? (In other words, how would you restrict the *domain*?)

Q8 We say that a rectangle's area *varies directly with* (or *is directly proportional to*) its height when its base is held constant. Describe in your own words what you think this means.

To answer Q10, compare a rectangle that produces a steep graph to one that produces a gradual graph.

Q9 Erase the traces. Drag point *B* to change the length of the base of the rectangle. What effect does this have on the graph?

Q10 How is the length of the base related to the slope of the graph?

EXPLORE MORE

Q11 Go to page 2 of **Direct Variation.gsp.** This page shows square *ABCD*. Does the area of the square vary directly with the length of one side? Make a prediction and then drag the points to check your prediction. Describe the path of the plotted point. Is this direct variation? Why or why not?

Inverse Variation

PUNCTUALITY

Q1 Yes, this is an inverse variation. Speed and time are variables, and the distance is a constant. In this context, distance is the distance from Adrianne's house to her school. That distance cannot change, so it is reasonable to call it a constant.

Q2 At 6:30, the time remaining (x) is 1.5 hours. The distance is 3 miles. Solving for y, Adrianne's average speed must be 2 mi/h.

$x = 1.5, y = 2$

Q3 She must walk 3 mi/h.

$x = 1, y = 3$

2. Check to see that everyone is plotting the first two points correctly, $(1.5, 2)$ and $(1, 3)$. After that, students should have no trouble plotting points on their own.

Q4 She has 0.75 hour remaining. She must travel 4 mi/h.

$(0.75, 4)$

Q5 She will have 0.5 hour remaining. She must ride her bicycle at 6 mi/h.

$(0.5, 6)$

Q6 She will have 1/3 hour remaining. The bus must travel 9 mi/h.

$(1/3, 9)$

It is possible to enter the fraction 1/3 in the Plot Points dialog box, or students can enter a decimal approximation.

Q7 There will be 1/6 hour remaining. Pete must drive 18 mi/h.

$(1/6, 18)$

Eighteen mi/h may sound slow for a car. Remind students that this is the average speed. Pete has to allow for traffic lights and stop signs, and he will have to find a parking space.

Q8 Answers will vary. Some students will observe that as one of the variables increases, the other decreases. Other students may be more precise, pointing out that the required speed is proportional to the multiplicative inverse of the remaining time.

GRAPH THE CURVE

Q9 $y = \dfrac{3}{x}$

Students may have already written the equation in this form. It would make it easier to answer the previous questions.

Q10 Since the curve represents the general solution for any given x, all of the plotted points should lie on the curve.

Q11 It is not possible for any two of the curves to intersect. If they did, the intersection point (x, y) would satisfy both equations. That means that the time and speed would be the same in both cases. If the time and speed are the same, then the distance traveled must be the same, but each curve represents a different travel distance.

This activity investigates the properties of an indirect variation in this form:

$$xy = k, \text{ where } k \text{ is a constant}$$

Tell students to imagine that they live 3 miles from school and that they must arrive there by 8:00. If they know how much time they have when they leave the house, they can compute what their average speed must be in order to arrive on time.

1. Write this formula: *speed · time = distance*

Q1 Ask students which number is a constant. It must be the distance because the distance between home and school does not change.

2. Define variables and units, and write them on the board. Write the equation too.

$$x = \text{time remaining before 8:00 (hours)}$$
$$y = \text{speed (mi/h)}$$
$$xy = 3$$

3. Draw the following table, with only the top row filled in:

	6:30	7:00	7:15	7:30	7:40	7:50
x	1.5	1	0.75	0.5	0.33	0.17
y	2	3	4	6	9	18

4. Model the first column for them. It is 6:30, so she has 1.5 hours to get to class. Substitute 1.5 for x in the equation, and solve for y.

5. Have students fill in the rest of the table.

Tell them to stay with the given units—no minutes or seconds.

Q2 Ask students to explain why this relationship is called an inverse variation. (As x becomes smaller, y grows larger, and vice versa.)

6. Open **Inverse Variation Present.gsp.** There is a set of coordinate axes at an appropriate scale. Ignore the buttons for now.

7. Choose **Graph | Plot Points.** Enter the (x, y) coordinates from the table.

Q3 What is the equation for y as a function of x? $\left[y = \frac{3}{x} \right]$

8. Press the *Show Graph* button to reveal the graph of the equation. Confirm that all of the plotted points are on the curve.

Q4 Challenge students to describe what the general shape of the graph would be if the distance were 1 mile. (Same, but closer to the axes.) What if it were 5 miles?

9. Press the *Show Other Curves* button to see the graph for 1, 2, 4, and 5 miles.

Q5 Is it possible for any of these graphs to intersect? Discuss.

Inverse Variation

Two variables, x and y, have an *inverse relationship* if y depends on the inverse of x:

$$y = \frac{k}{x}, \text{ where } k \text{ is a constant}$$

You can also express this in the form $xy = k$: The product of x and y is constant.

PUNCTUALITY

Adrianne has been late for school twice this week. She has resolved to become more punctual (and avoid detention). Her home is 3 miles from school, and the first class starts at 8:00. She figures that if she leaves the house at 6:30, she can walk to school at a leisurely pace and still arrive well before the bell. Before leaving, Adrianne does a quick calculation to see how fast she needs to walk. For this, she uses a formula that relates speed, time, and distance:

$$speed \cdot time = distance$$

Q1 Is this an inverse variation? Which numbers are the variables, and which is the constant? Is it reasonable to call that number a constant?

Let the *x* units be hours, and let the *y* units be mi/h. Do not use any other units.

Q2 Let x be the time (in hours) remaining before class, and let y be the speed that Adrianne needs to travel in order to arrive at 8:00. It is now 6:30. What is x and what is y? How fast must Adrianne walk?

Q3 While Adrianne was looking up the formula, 30 minutes passed. It is now 7:00. Compute x and y again. How fast must she walk if she leaves now?

It will have to be a fast walk. She decides to graph it, just to be thorough.

1. In a new sketch, choose **Graph | Grid Form | Rectangular Grid.** The rectangular grid allows you to adjust the x and y scales independently.

2. Choose **Graph | Plot Points.** Plot the (x, y) pairs that you computed in Q2 and Q3.

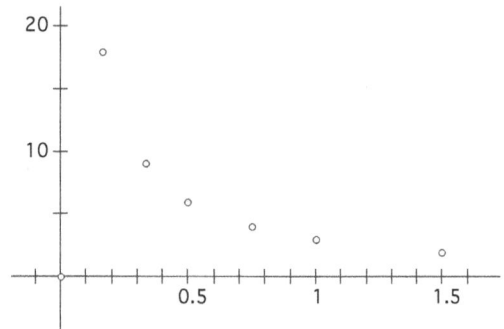

In the Plot Points dialog box, you can enter simple expressions. For example, enter 1/3 for one third of an hour (20 minutes).

Q4 The graph took longer than she expected. Now it's 7:15. How much time does she have left? How fast must she travel? Plot the point.

Q5 Now Adrianne has decided that it would be a better idea to ride her bicycle. There is still plenty of time if she leaves now (OK, as soon as this TV show is over). How fast will she have to ride if she leaves at 7:30? Plot the point.

Q6 Wouldn't you know it? Now it's raining. She is absolutely not going to school with wet hair. The bus will come by at about 7:40. How fast will it have to travel to get Adrianne to school on time? Plot the point.

Q7 Adrianne must have dozed. She's missed the bus. Now she will have to ride to school with her brother, Pete. He always drives too fast. He doesn't even leave the house until 7:50. How fast will he have to drive? Plot the point.

Q8 Look at the values you have found so far. In your own words, explain why the relationship between these two quantities is called *inverse variation*.

It's 7:50 now. Where's Pete? Adrianne's mother tells her that Pete took the bus. His car broke down yesterday. This is just too much! Now she'll have three tardy notices in one week, and it wasn't even her fault.

GRAPH THE CURVE

Adrianne now realizes that she was wasting too much time on calculations. Rather than compute each speed separately, she should have written y as a function of x. She could then plot the curve and find the speed for a whole range of departure times.

Q9 Write an expression for y in terms of x.

3. Choose **Graph | Plot New Function.** For the function definition, enter the formula for y from your answer to Q9.

Q10 What is the relationship between this curve and the points you plotted earlier?

To change the domain of a graph, select the graph and choose **Edit | Properties | Plot.** Negative time would make no sense here, so let zero be the minimum value of x.

4. This graph applies only to students who live 3 miles from school. Draw similar graphs for students living 1, 2, 4, and 5 miles from school. Use different colors. Label the axes and the curves.

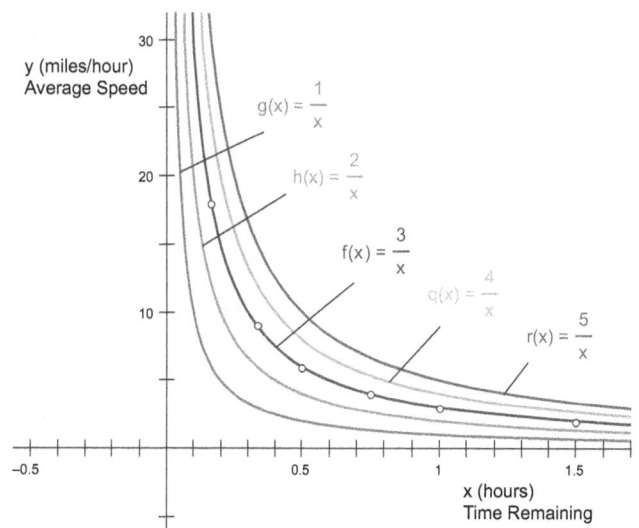

Q11 Do any of these curves intersect? Explain why or why not.

Modeling Projectile Motion

SKETCH

Q1 During the first period of time, the velocity's magnitude is reduced and its direction is less steeply upward. Students may not be familiar with the terms *magnitude* and *direction*. It's not important that they learn the terms at this time, but it is important that they recognize both changes: The ball is going more slowly, and its direction has changed so that its path is less steep than it was at first. Students can see both effects by comparing vectors v_0 and v_1. Vector v_1 is shorter in length and is pointed less steeply upward.

Q2 With each interval of time, the velocity vector points more downward. As the velocity vector moves downward, the ball is not moving upward as steeply as it was at first (and it soon starts to move downward).

Q3 The path of the ball levels off, becoming more horizontal and less vertical. Soon the ball begins to move downward. (In the default state of the sketch, the ball reaches its apex near p_3 and then begins to move downward.)

Q4 If the velocity pattern continues, the ball's direction of travel will be more and more downward at each time interval. This agrees with students' knowledge of how balls fly through the air, but it is also important that they observe how the changes in the velocity vector correspond to the actual movement of the ball.

Q5 The shape of the ball's path looks like an arc of a parabola. The more downward the velocity vector points, and the longer it gets in this direction, the more vertical the ball's path becomes.

Q6 In the original state of the sketch, the ball begins to move downward during the fourth interval. The v_3 vector is the first one to point downward, with the result that after p_3, the ball begins to move downward.

Q7 Answers will vary. If there is no gravity, the velocity vector will not change over time. With no changes in velocity, the ball will continue to travel in the initial direction, and its path will be a straight line. Encourage students to explain their drawings, and make sure they understand that the ball moves in a straight line forever because gravity does not alter the ball's velocity.

Q8 Students have a mental image of how a thrown ball moves through the air, but it's still useful to demonstrate by gently tossing a soft object.

The real ball moves through the air in a smooth arc, while the construction is a rough approximation because its position is only measured at intervals, where the segments connect.

Some students may raise a more sophisticated question about the terminal velocity of a ball and point out that the Sketchpad model doesn't allow for air resistance.

Q9 You should shorten both the velocity and gravity vectors by half. You should shorten the velocity vector because the ball should travel less far in a shorter interval of time. Similarly, you should shorten the gravity vector because during a half-second interval the ball's velocity should not change by as much as it does during a full-second interval.

Q10 As it gets smoother, the iterated path becomes more and more parabolic in shape.

EXPLORE MORE

10. Students can also make the parabola and the iterated path match by changing the path. This approach requires modifying the initial position of the ball, the initial velocity of the ball, and the length of the gravity vector. (The gravity vector must remain vertical or the objects won't match.)

Modeling Projectile Motion

When you toss a ball into the air, the laws of physics control its motion through the air and back to the ground. In this activity you'll use Sketchpad's **Iterate** command to model a projectile as its position, velocity, and gravity interact.

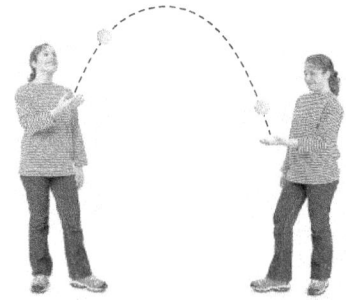

To create the model, you'll make a sketch showing what happens to the projectile in a short period of time. You'll use iteration to repeat the construction and create a graph of the projectile's motion over a much longer period of time.

SKETCH

1. Open **Projectile Motion.gsp.** Notice the vector for gravity (g), the vector for the initial velocity (v_0) of the object, and the point for the initial position (p_0) of the object. Vector g points down to represent the downward acceleration gravity exerts on projectiles. Vector v_0 points diagonally to indicate the movement of the ball at the instant it is thrown.

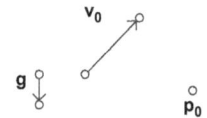

At this instant, the ball has not moved from its initial position. To begin, you'll construct a segment to represent the ball's motion by using the velocity vector to find the ball's position after the first interval of time.

> In the Translate dialog box, select **Marked** for all translations.

2. To mark the velocity vector as the vector for translation, select its initial and final points, and choose **Transform | Mark Vector.** To translate the position, select point p_0 and choose **Transform | Translate.** Label the translated position p_1 and connect the initial position and the translated position with a segment.

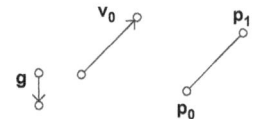

The new segment shows how the ball moves through the air during the first period of time. But if you've ever thrown a ball, you know that gravity affects its flight. In your sketch you must also take into account gravity's effect on the ball's velocity. To do this, you'll use the gravity vector to construct a new velocity vector.

> This segment represents the velocity vector at the end of the first period of time.

3. Mark the gravity vector as the vector for translation. Translate point v_0 by this vector, and then construct a segment from the original tail of the velocity vector to v_1 (the new translated point).

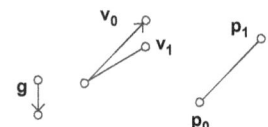

The v_1 vector represents the object's velocity at the end of the first interval of time. So v_1 describes the ball's velocity when it's at p_1.

New York City Title I High School Activities with The Geometer's Sketchpad
© 2012 Key Curriculum Press

4. Repeat steps 2 and 3 for the second period of time. Be sure you construct the new position segment from the end of the old one, but the new velocity vector with its tail in the same place as the existing velocity vectors.

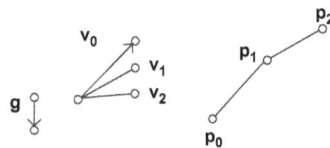

Q1 How did the velocity change as the ball moved from p_0 to p_1? Is the ball going faster or slower? How did the direction of its motion change?

5. Repeat steps 2 and 3 for a third period of time.

Q2 What do you notice when you compare the velocities v_0, v_1, v_2, and v_3 for the three periods? How is the velocity vector changing the object's path?

Q3 What do you notice when you compare positions p_0, p_1, p_2, and p_3? Describe the motion of the object.

Q4 Make a prediction for how the position will change during the next ten periods of time.

It would be a lot of work to do the same construction ten more times. You can use iteration to make the process easier.

Choose **Edit | Undo** repeatedly until you get back to the end of the first period.

6. Undo your work back to the end of the first period of time.

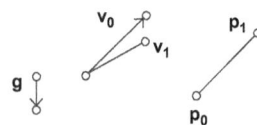

Gravity is constant, so you didn't translate that vector. This means that you only have to keep track of changes in the position and the velocity. The first step of the iteration will map $p_0 \to p_1$ and $v_0 \to v_1$.

Points p_0 and v_0 are called the *pre-image* points of the iteration, and points p_1 and v_1 are the *image* points.

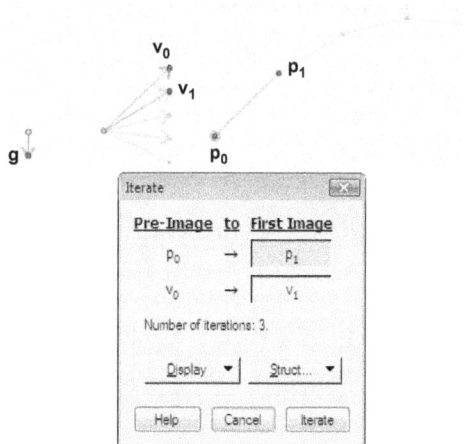

7. To create the iteration, select the pre-image points p_0 and v_0 and choose **Transform | Iterate**. Then click p_1 and v_1 in the sketch so that each pre-image point is correctly mapped to its image point. Click **Iterate** to complete the iteration and show the next three steps.

8. With the iterated images still selected, press the **+** key on your keyboard to increase the number of iterations until you have about 15 of them.

Q5 Describe the shape of the object's flight. For each period of time, how do the direction of the velocity vector and the path of the object relate to each other?

Q6 During which period of time does the velocity begin to point downward? Can you tell at what moment the ball begins to fall downward by looking only at the velocity vector? Can you tell when the velocity vector begins to point downward just by looking at your construction of the ball's path?

To turn gravity off and check your prediction, drag the end points of the gravity vector so that they coincide.

Q7 What do you think the path would look like if there were no gravity? Draw your prediction on paper.

Q8 How well does the shape of the path represent the actual flight of a ball through the air? Think of a time when you've seen a ball fly through the air. How does your constructed path differ from your observation of how balls fly through the air?

The model can be more accurate if you make each iteration represent a shorter period of time. To make this change, you will have to change the lengths of the original vectors.

Q9 To make each period of time half of its original value, how much shorter must you make the vectors? How many more iterations will you need to model the same period of time?

9. Shorten the gravity and initial velocity vectors and increase the number of iterations until your model shows smooth motion for the projectile.

Q10 What shape does the projectile's motion appear to have?

EXPLORE MORE

10. Using **Graph | Plot New Function** and the form $f(x) = ax^2 + bx + c$, try to match a parabola to the path of the ball. Adjust the parameters to change the shape of your parabola.

11. Extend the model to use a parameter to represent the time interval. (*Hint:* Use the parameter to dilate the original vectors before you iterate them.) Page 2 of the sketch describes one way to do this.

PROPERTIES OF THE GRAPH

Q1 Both a and b are nonzero. Therefore, ab^x is also nonzero for any real x, so y cannot be zero, and there is no x-intercept.

For the y-intercept, substitute zero for x in the equation $y = ab^x$.

$$y = ab^0 = a$$

The y-intercept is a.

Q2 For $b > 1$, $f(x)$ tends to zero on the left. For $0 < b < 1$, $f(x)$ tends to zero on the right. The value of a has no influence on this property.

Q3 If a were equal to zero, the function would be the constant function $f(x) = 0$.

If b were equal to zero, the function would be zero for all positive x, and it would be undefined for all other values of x.

If b were less than zero, b^x would not be continuously defined over all real exponents x.

If b were equal to one, this would be another constant function, $f(x) = a$.

Q4 Limited resolution may prevent students from making the difference exactly 1.00. It's sufficient for them to make it as close to that value as they can. The ratio $y_Q/y_P = 1.30$. This is the same as parameter b.

$$\frac{y_Q}{y_P} = \frac{f(x_Q)}{f(x_P)} = \frac{ab^{x_Q}}{ab^{x_P}} = b^{x_Q - x_P} = b^1$$

Here it does not matter where on the graph points P and Q are, so long as $x_Q - x_P = 1$.

Q5 If $x_Q - x_P = 2.5$, then $y_Q/y_P = 1.30^{2.5} \approx 1.9$. This follows from the same reasoning as in the previous answer.

$$\frac{y_Q}{y_P} = \frac{f(x_Q)}{f(x_P)} = \frac{ab^{x_Q}}{ab^{x_P}} = b^{x_Q - x_P} = b^{2.5}$$

DOUBLING PERIOD AND HALF-LIFE

8. There may be some confusion regarding the term *effective annual yield*. The actual rate is about 5.8%, but since it is compounded continuously, at the end of each year, the investment will be worth 6% more than it

was at the beginning of the year. Even if students do not yet grasp the concept, they can continue with the given function definition.

Q6 The doubling period is about 11.90 years.

Q7 To find the half-life, students should arrange the points so that $y_Q/y_P = 0.50$. When that happens, $x_Q - x_P \approx 30$, so cesium has a half-life of about 30 years. In this case changes in the scales of the axes can cause quite a lot of variation in the answers.

Q8 To find a doubling period, with the ratio $y_Q/y_P = 2.00$, point Q would have to be to the left of P. In that case, $x_Q - x_P \approx -30$.

This answer fits with the half-life answer. It stands to reason that if it takes 30 years for half of the cesium to decay, then 30 years ago, there was two times as much. This explains the negative doubling period.

Exponential Functions

1. Begin by showing the general form of this exponential function:

$$f(x) = ab^x, \quad \text{where } a \neq 0,\ b > 0,\ \text{and } b \neq 1$$

Press *Show Slider Controls* to reveal buttons that you can use to set the parameters to various precise values.

2. Open **Exponential Present.gsp.** This is a graph of the function with sliders controlling the values of a and b. Drag each slider in turn so that students can see the effects on the graph.

Q1 Ask students for the x- and y-intercepts. [There is no x-intercept, and the y-intercept is equal to a.] Challenge them to verify these facts by alternately setting y and x to zero in the equation $y = ab^x$.

Q2 Sometimes the function approaches zero on the left side, and sometimes on the right. What determines which side it is? [It's on the left when $b > 1$ and on the right when $0 < b < 1$.]

Q3 In the general form of the function, there are three restrictions on the parameters a and b. Why? Show students what happens when $a = 0$, $b < 0$, $b = 0$, or $b = 1$.

3. On page 2 there are two points on the graph, P and Q. You can drag P freely, but point Q is a fixed distance to the right of point P. That distance is determined by the slider labeled Δ. At the bottom of the screen are measurements showing the difference of the x-coordinates and the ratio of the y-coordinates.

4. Drag point P to show that the ratio of the y-coordinates remains constant when the difference in the x-coordinates (Δ) is constant.

On pages 3 and 4, the a and b sliders have been replaced with parameters in order to make it easier to enter precise values.

Q4 What will the ratio be when $\Delta = 1.00$? [It should equal b.] Challenge students to predict this before showing it. Then have them prove that $\dfrac{f(x_P + 1)}{f(x_P)} = b$.

5. Page 3 has a graph showing the growth of an investment of $100 with an effective annual yield of 6%.

Pages 3 and 4 have rectangular grids. If students have not used that feature yet, this would be a good opportunity to show them the advantages of using different scales on the axes.

Q5 With this investment, how long would it take to double your money? If the money is doubled between P and Q, then the ratio of their y-coordinates will be 2. Drag the Δ slider until the ratio is 2.00. [The doubling period is the difference in the x-coordinates, about 11.9 years.]

6. The graph on page 4 shows the radioactive decay of 80 g of cesium. The x-scale is in years.

Q6 What is the half-life of cesium? This question is similar to the previous one. Give students time to figure out that they need to adjust the difference in the x-coordinates so that the ratio is 0.50. [The result is a half-life of about 30 years.]

Exponential Functions

There is a connection between population growth, radioactive decay, musical scales, and compound interest. They seem to have little in common, but you can model any of them using an exponential function.

An exponential function has the general form $f(x) = ab^x$, where $a \neq 0$, $b > 0$, and $b \neq 1$.

PROPERTIES OF THE GRAPH

Before using an exponential function to model a real-world problem, take some time to familiarize yourself with the graph.

> To create a parameter, choose **Number | New Parameter.**

1. In a new sketch, create parameters a and b.

2. Graph the function $f(x) = a \cdot b^x$ by choosing **Graph | Plot New Function.** Click the parameters in the sketch to enter them into the function.

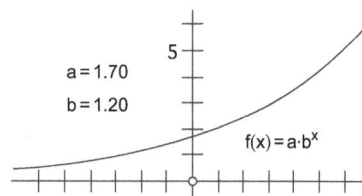

$a = 1.70$
$b = 1.20$
$f(x) = a \cdot b^x$

> To change the value of a parameter, either double-click it or select it and press the + or − key. (To change the size of the steps for the + or − keys, select the parameter and choose **Edit | Properties | Parameter.** Change the Keyboard Adjustments value to 0.1 unit.)

3. The graph is plotted on the screen. Change the values of the parameters, and observe the resulting changes in the graph. Try several values for each parameter.

Q1 What are the x- and y-intercepts of the graph? Explain how the intercepts are related to parameters a and b.

Q2 The value of $f(x)$ tends to get close to zero either on the left side or on the right. What parameter values determine which side it is?

Q3 In the general form of the exponential function, there are three constraints ($a \neq 0$, $b > 0$, and $b \neq 1$). Explain the reason for each of these constraints.

Next you'll investigate how the function behaves by comparing the coordinates of two points on the graph.

4. Change the parameters so that $a = 2.00$ and $b = 1.30$.

5. Construct two points on the function graph. Label them P and Q.

6. Select both points and choose **Measure | Abscissae (x).** Select the points again and choose **Measure | Ordinates (y).**

7. Calculate the values $x_Q - x_P$ and y_Q / y_P.

$x_P = -1.38$ $y_P = 1.71$
$x_Q = 2.55$ $y_Q = 3.50$
$x_Q - x_P = 3.94$
$\dfrac{y_Q}{y_P} = 2.05$

> You can change the scale of the axes to give you more precise control over the positions of the points.

Q4 Drag point Q one unit to the right of P, so that the difference $x_Q - x_P$ is as close to 1.00 as you can make it. What is the value of the ratio y_Q / y_P? Drag point P to a different

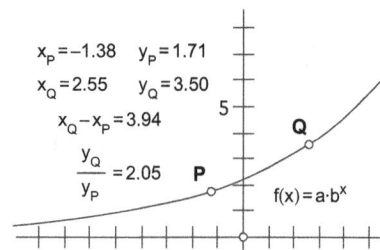

position on the graph, and again drag Q so it's one unit to the right. What is the value of the ratio y_Q/y_P? Why do you get this result?

Q5 Now use a difference other than 1. Drag the points so that $x_Q - x_P$ is approximately 2.5. What is the ratio? Drag them to another position, but with the x difference still equal to 2.5. What is the ratio now? Explain.

DOUBLING PERIOD AND HALF-LIFE

Exponential functions can be used to solve a number of real-life problems. First change the function so that it shows the value of $100 invested at an effective annual yield of 6%.

Effective annual yield is not the same thing as interest rate. That's another topic.

8. Edit parameters a and b so that the function has the definition $f(x) = 100(1.06)^x$. This shows the value of $100 invested at an effective annual yield of 6%. (The x variable is in years, and y is in dollars.)

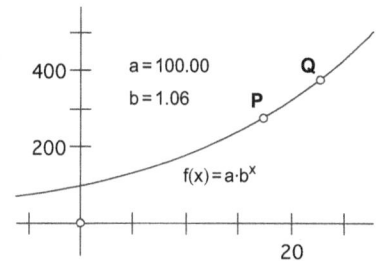

Setting the grid to rectangular allows you to adjust each axis independently of the other.

9. At first you can't see the graph because the y-axis doesn't go up to 100. To adjust the axes, choose **Graph | Grid Form | Rectangular Grid.** Then drag tick mark numbers on each axis so that you can see the results for the first 25 years.

To see more decimal places in a parameter, select the parameter and choose **Edit | Properties | Value.** Change the Precision setting.

Q6 How long will it take to double your money? Drag the points so that the ratio is 2.00. What is the difference in their x-coordinates? This number is called the *doubling period.*

An exponential function can also be used to model the decay of radioactive cesium.

10. To model the decay of 80 g of cesium, change the function definition to $f(x) = 80(0.977)^x$. Adjust the axes appropriately. The value of x is still in years, but the value of y is in grams.

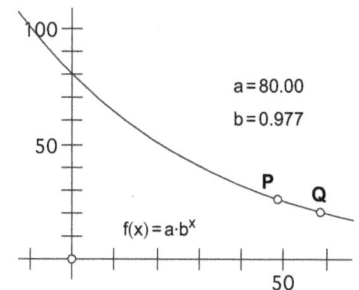

Q7 If you start with 80 g, you will have less cesium every year. How long would it take to lose half of it? Explain how you found the answer. This number is called the *half-life* of cesium.

Q8 Although cesium decays, as opposed to growing, you can still calculate its doubling period. Drag the two points until you find a position where the ratio is 2.00. What is the difference in the x-coordinates? Explain how this verifies your answer to Q7.

Logarithmic Functions

GRAPH INVERSE EXPONENTIAL FUNCTIONS

On page 1 students graph $y = 2^x$ to begin with an exponential function that has points whose coordinates they can verify in their heads. On page 2 they use a base of 10 to make it easy to see that 10^x and $\log x$ are inverse functions. On page 3 they use function transformations to find the inverse of a more general exponential function, and end up with a conjecture as to how to write the formula for such a function. This portion of the activity is useful either before introducing the change-of-base property of logarithms (to motivate that method) or afterward (to review the method).

2. When students press *Show Points*, the seven points are shown and selected. Because the points are selected in order, it's easy for students to measure and then tabulate the coordinates. Be sure they measure the coordinates using **Measure | Coordinates** rather than measuring the x- and y-values separately. If they deselect all objects at any point in the process, they can press *Show Points* to select objects again in order.

Q1 A negative value of x in the exponential function means that the exponent is negative. A positive base raised to a negative exponent results in a positive number.

Q2 Within each pair of points (A and A', B and B', and so forth), the x- and y-values are interchanged.

Q3 None of the x-coordinates of the reflected points can be negative, because none of the y-coordinates of the original function were negative.

Q4 The line $y = x$ is called the *axis of symmetry* because it lies precisely between the two graphs, reflecting each onto the other.

Q5 The graph of the log function coincides with the reflected image of the exponential graph, so the log function must be the inverse of the original exponential function.

13. This instruction is brief and does not give precise details concerning commands. If students have trouble with it, have them review what they did in steps 7–8.

Q6 The two graphs match when a is approximately 3.32 and $b = 1$. This value of $a = 3.32$ is approximately $1/\log 2$, but students don't have enough information yet to reach this conclusion.

Q7 When k is 10^1, $a = 1$; when k is 10^2, $a = 1/2$; when k is 10^3, $a = 1/3$. It appears that a must be $1/\log k$.

New York City Title I High School Activities with The Geometer's Sketchpad
© 2012 Key Curriculum Press

Q8 Because $10{,}000 = 10^4$, the value of a must be $1/4$. Because $0.1 = 10^{-1}$, the value of a must be $1/-1 = -1$. Students should verify these predictions by testing them in the sketch.

Q9 The inverse of $f(x) = k^x$ is $f^{-1}(x) = \log x / \log k$.

EXPLORE MORE

Q10 Define $d = \log k$. Rewrite k as 10^d. Then:

$$y = k^x = \left(10^{\,d}\right)^x = 10^{dx}$$

$$\text{so } \log y = dx$$

$$\text{and } x = \frac{\log y}{d} = \frac{\log y}{\log k}$$

Q11 The inverse function is $f^{-1}(x) = b\log((x-k)/a) + h$

In this presentation students will find the log function that's the inverse of $f(x) = k^x$.

Remind students that
the ordered pairs of an
inverse function are in
the opposite order.

1. Open **Logarithmic Functions Present.gsp.** Press *Show Exponential Function.* The graph of $f(x) = 2^x$ appears. Press *Show Points* and then *Show Coordinates* to show seven points on the graph, along with their coordinates.

   Q1 On the inverse function, what are the coordinates of the point corresponding to *A* on the original function? (8.00, 3.00) Ask different students to give coordinates for points on the inverse corresponding to each of the other points.

   Q2 What geometric transformation do you know that switches the *x*- and *y*-values? [reflection across the line $y = x$]

2. Use the next three buttons to reflect the points across $y = x$ and show the coordinates. The two tables confirm the answers students gave for Q1.

3. To see the entire graph of the inverse function, press *Show Reflected Graph.*

Now compare an exponential function with a logarithmic function, and demonstrate that they are inverses by verifying that each is a reflection of the other across $y = x$.

4. On page 2, use the top two buttons to show the exponential function $f(x) = 10^x$ and the logarithmic function $g(x) = \log x$.

5. Use the next two buttons to show $y = x$, a point on $f(x)$, and its reflection.

6. Use the Animate button to animate the point, trace the reflection, and verify that the reflection (and therefore the inverse) of $f(x) = 10^x$ really is $g(x) = \log x$.

Now graph an exponential function with an adjustable base, and graph a stretchable logarithmic function. Match the logarithmic function to the inverse of the exponential function, and find a formula for the inverse of the exponential function.

7. On page 3, press the first three buttons to show the function $f(x) = k^x$ and its inverse. Use the green buttons to switch *k* from 2 to 5 and back to 2 again.

8. Show the adjustable logarithmic function, and drag sliders *a* and *b* to make the red logarithmic function match the blue inverse function.

9. Show the table of values for *k, a,* and *b.* Double-click the table to make the current values permanent.

10. Set *k* to the values 5, 10, 100, and 1000. For each value of *k*, drag sliders *a* and *b* to match the inverse, and record the values permanently in the table.

    Q3 Ask, "How can you determine the values of *a* and *b* from *k*?" [The value of *b* is always 1, and the value of *a* is 1/log *k*.]

11. Test this conclusion by using the remaining green buttons to adjust *k*, predicting the required values of *a* and *b*, and testing the predictions by dragging *a* and *b*.

Logarithmic Functions

Many occurrences in our natural world can be modeled using logarithmic functions, including the strength of earthquakes, the intensity of sound, or the concentration of hydronium ions in a solution. In this activity you'll explore the relationship between exponential and logarithmic functions, and determine how to write the formula for a logarithmic function that's the inverse of a particular exponential function.

GRAPH INVERSE EXPONENTIAL FUNCTIONS

Logarithms are related to exponents, so start by graphing an exponential function and finding the inverse graph.

1. Open **Logarithmic Functions.gsp.** Press the *Show Exponential Function* button to see the exponential function $y = 2^x$ along with its graph.

With the points selected, choose **Measure | Coordinates** to find their ordered pairs. With the coordinates selected, choose **Number | Tabulate** to place the coordinates in a table.

2. Press the *Show Points* button to show seven points on the curve. Measure their coordinates, and put the resulting ordered pairs into a table.

Q1 Notice that some of the *x*-values are negative. Does this mean that the resulting values of the function are negative? Explain why this is true or not true.

Next, interchange the *x*- and *y*-values by reflecting the points over the line $y = x$.

3. Press the *Show y=x Line* button. With the line selected, choose **Transform | Mark Mirror.** The line flashes briefly to indicate that it is marked as the mirror.

To show the labels of the selected points, choose **Display | Show Labels.**

4. Press *Show Points* again to select the seven points in order, and choose **Transform | Reflect.** The seven points are reflected. Show their labels.

5. Measure the coordinates of the reflected points and tabulate the results. Align the two tables in order to see the original and reflected points next to each other.

Q2 What do you notice about the coordinates of each pair of points?

Q3 Will any of the *x*-coordinates of the reflected points be negative? Explain.

Q4 Why is the line $y = x$ called the *axis of symmetry* for a function and its inverse?

Next reflect the entire graph over the line $y = x$.

Use the **Point** tool to construct the new point.

6. Construct a point on the original graph, and reflect it over the line $y = x$. Drag your new point and observe the behavior of the reflected image.

The reflected graph is the graph of the inverse of the original function.

7. To create the entire reflected graph, select the point on the graph and its reflected image, and choose **Construct | Locus.** Change the color of the locus, and make it dashed.

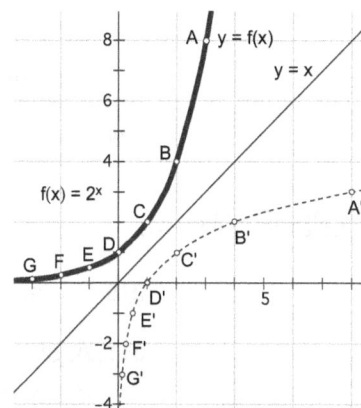

You'll use page 2 to graph $y = 10^x$, reflect it to show its inverse, and compare the inverse to the graph of $y = \log x$.

8. On page 2 construct the graph of $y = 10^x$ by choosing **Graph | Plot New Function** and entering 10^x into the Calculator.

9. Construct a point on the graph and reflect it across the graph of $y = x$.

*To turn tracing on or off, select the point and choose **Display | Trace Point**.*

10. Turn on tracing for the reflected point, and drag the point on the graph to observe the shape of the inverse function.

11. Construct the graph of $y = \log x$.

Q5 What do you observe about the graph of the log function and the reflected image of the exponential graph? What conclusion can you draw?

On page 3 you'll graph the exponential function $f(x) = k^x$ and use different values for k to find a general formula for the logarithmic function that's the inverse of $f(x)$.

*Choose **Number | New Parameter** to create the new parameter, and set its label to k and its value to 2 in the dialog box that appears.*

12. Create a parameter k, set its value to 2, and use it to construct the graph of $y = k^x$. To enter k into the function definition, click its value in the sketch.

13. Construct a point on the graph, reflect it across $y = x$, and construct the locus.

This locus is the graph of the inverse function. Express this inverse as a logarithmic function by stretching or shrinking the parent logarithmic function $y = \log x$.

14. Using the values of a and b, plot the logarithmic function $y = a\log(x/b)$. Adjust the sliders so that your newly plotted function matches the inverse of $y = k^x$.

Q6 What values of a and b made the graphs match?

To add a row to a table, making the current values permanent, double-click the table.

15. Record the values of k, a, and b in a table. Then change the value of k to 5, match the graphs again, and add a new row of data to the table. Continue adding new data to the table for the following values of k: 10, 100, and 1000.

Q7 What pattern can you find to relate the value of k to the values of a and b?

Q8 Use this pattern to predict the values of a and b needed when $k = 10,000$. Test your prediction by gathering another row of data for your table. Then predict the values needed when $k = 0.1$, and test your prediction.

Q9 Use your results to write a formula for the inverse of $f(x) = k^x$.

EXPLORE MORE

Q10 Use algebraic manipulation to explain why your formula from Q9 must be true.

Q11 A general exponential function can be written as $f(x) = a \cdot 10^{(x-h)/b} + k$. Write the corresponding inverse function in terms of a, b, h, and k.

It's a good idea to have an initial class discussion about domain and range before starting the activity, and it's very important to have a class discussion after students finish the activity.

One or the other of these discussions should focus on why some numbers are not allowed as inputs for particular functions. Ask students to give examples of functions with restricted domains. One category of answers is covered in the activity—functions such as square root functions and rational functions that have undefined outputs for certain inputs. Another category, not covered in the activity, is functions that model real-life or geometric situations. For example, the function $f(n) = 0.89n$ might represent the total cost of buying n apples each costing \$0.89. But it doesn't make sense to consider $n = -2$ or $n = 3.71$ here. Only positive whole numbers are part of this domain.

A good way to encourage exploration is to ask students to modify the various functions and observe the results. On page 1, you could ask students how the range would change if they changed the function to $f(x) = \text{round}(x) + 1$ (*answer:* it wouldn't). You could then ask for what values of k would the function $f(x) = \text{round}(x) + k$ have a range different from that of the original function (*answer:* non-integer values). You could follow this up by exploring what happens when you multiply $\text{round}(x)$ by a constant. You could extend the activity in similar ways on the other pages.

SKETCH AND INVESTIGATE

Q1 $g(x)$: all real numbers
$h(x)$: $h(x) \geq 0$
$j(x)$: all real numbers
$k(x)$: $-2 \leq k(x) \leq 2$

Q2 The range appears to be all integers, as with $f(x) = \text{round}(x)$. This is clearly wrong since many inputs result in non-integer outputs. For example, $g(0.01) = 0.2$.

The output marker lands only on integers because the input marker can't really be dragged continuously. You can only move it by one pixel (screen unit) at a time. This dynagraph's scale is set to 1 pixel = 0.05 units. Moving the input marker by a single pixel moves the output marker by 20 times as much, which is equivalent to a full unit. You can change the scale by pressing the *Show Scale* button and adjusting the slider.

In the next step, students are encouraged to use animation rather than dragging the marker directly, because an object being animated can be slowed down so that it moves less than a pixel at a time.

Q3 The domain of function $v(a)$ is $a \geq 0$. (When the input marker is to the left of 0, the output marker disappears.)

The range is $v(a) \geq 0$.

Q4 Negative numbers are not in the domain because the square root of any negative number is undefined over the set of real numbers.

Q5 The domain restriction is $b \neq 2$, and the range restriction is $w(b) \neq 0$. The value 2 is excluded from the domain because it would result in division by 0. The value 0 isn't part of the range because the result of the division $\frac{1}{x-2}$ cannot be 0.

DOMAIN AND RANGE ON CARTESIAN GRAPHS

On page 5, the graph's properties are set to plot it discretely rather than continuously. If the plot were done continuously, the discontinuities would be connected with segments. To change whether a graph is plotted continuously or discretely, select the object and choose **Edit | Properties | Plot.**

Q6 One way to determine domain by looking at a Cartesian graph of a function is to imagine a vertical line that sweeps from left to right. Any location where the line touches the graph is part of the domain. Thus, if a vertical line crossing the x-axis at $x = 3$ touches the graph somewhere, 3 is part of the domain of that function. Any location where the line doesn't touch the graph at all is not part of the domain.

Similarly, to determine the range from a Cartesian graph, imagine a horizontal line sweeping from bottom to top. Anywhere it touches some part of the graph is part of the range; anywhere it doesn't, isn't.

EXPLORE MORE

Q7 To make the range of f all even numbers, use $f(x) = 2 \cdot \text{round}(x)$.

To make its range all odd numbers, use $f(x) = 2 \cdot \text{round}(x) + 1$.

Q8 To make the domain and range of v all numbers less than or equal to 0, use $v(x) = -\sqrt{-x}$.

To make w's domain all real numbers except 0 and its range all real numbers except 2, use $w(x) = \frac{1}{x} + 2$.

WHOLE-CLASS PRESENTATION

Open **Domain Range Present.gsp** and use the pages of this sketch to stimulate a class discussion.

1. On page 1, drag input marker a and ask students to observe the possible positions of the input and output markers, and use their observations to describe the domain and range. After several students have volunteered descriptions in their own words, turn on tracing and drag again to verify the descriptions. You can use the Animation button to achieve smooth movement of the input marker.

2. For each of the four functions on page 2, ask students to guess ahead of time what the range will be. Then drag the input marker for that function to test their guesses. For each function, use tracing and animation to generate a smooth, detailed visual representation of the answer.

3. Use page 3 to emphasize the need to pay attention to details and not to jump to conclusions. By dragging the input marker, you'll generate what appears to be a range of integer values only. Ask students to explain what's going on here. This should generate a lively discussion. Encourage a number of students to describe the phenomenon in their own words. Finish this page by using animation to achieve movement by less than a pixel at a time.

4. On page 4, tracing shows the domain and range restrictions for $v(a)$ clearly, but cannot show the restrictions for $w(b)$ so clearly. Get students to discuss the differences in the two situations, so that they realize that $w(b)$ is missing only a single number in its domain and a different number in its range.

5. On page 5, drag input marker a and have students observe both the dynagraph and Cartesian graph. Ask them to explain how the restricted range shows up on the Cartesian graph. Also ask them how to edit the function to generate only even numbers, or only odd numbers.

6. Use page 6 to compare the dynagraph and Cartesian representations for the functions from page 2.

Finish by asking students to summarize what they learned about domain and range.

Domain and Range

You can't put a television in a blender, and you wouldn't expect an elephant to come out of a gasoline pump. In math terms, a television isn't an *allowable input* for a blender; it's not part of a blender's *domain*. And an elephant isn't a *possible output* of a gasoline pump; it's not part of a gas pump's *range*.

Similarly, functions have certain numbers that are and aren't allowed as inputs, and other numbers that are and aren't possible as outputs. In this activity you'll explore these notions using both dynagraphs and Cartesian graphs.

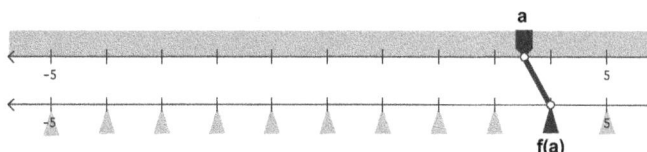

SKETCH AND INVESTIGATE

The *range* of a function is the set of possible outputs from that function. Let's see how dynagraphs can make this idea clearer. You'll start by exploring the range of everyone's favorite dynagraph: the "blue hopper."

1. Open **Domain Range.gsp.** Drag the input marker to observe the behavior of the function $f(x) = \text{round}(x)$. The input marker leaves a trace.

You can drag the input marker to any value you want, so we say "the domain of f is all real numbers."

*To turn on tracing, select the output marker and choose **Display | Trace Triangle.***

2. Turn on tracing for the output marker. Then drag the input marker back and forth again.

The output marker leaves a trace of where it's been. These traces point to every integer but never to any other values, so we say "the range of f is all integers."

The range of each function on page 2 will either be "all real numbers" or an inequality such as $f(x) \geq 5$.

Q1 On page 2, use the same method to find the range of each of the four functions.

When you use technology, it's very important to think about the limitations of that technology. You'll see that the method used above can be misleading in certain situations, and you'll then learn a more reliable method.

3. On page 3, turn on tracing for the output marker, then drag the input marker, as in step 2. For greater control, use the right and left arrow keys on your keyboard to drag one pixel (screen unit) at a time.

Q2 What does the range of $k(x) = 20 \cdot x$ *appear* to be? Explain why this answer is actually wrong. Why do you think this happens?

4. Select the input marker and choose **Display | Animate Pentagon.** Repeatedly press the *Decrease Speed* button (the down arrow on the Motion Controller) until it's clear that the range of this function really is all real numbers. Go back to pages 1 and 2 of the sketch, and convince yourself that your answers there were correct.

For most functions, the domain is "all real numbers," meaning that any input produces some output. Sometimes, however, a domain might be *restricted* to something such as "all integers" or "$x > 3$."

Q3 On page 4, drag v's input marker back and forth. What is v's domain? (In other words, where can you drag the input marker and still see the output marker?) What is v's range?

Q4 Based on its equation, why are some numbers not part of v's domain?

Q5 The domain of w is all real numbers except for one particular value. The range of w is also all real numbers except for one particular value (a different value). What is the one value not in the domain of w? What is the one value not in the range of w?

DOMAIN AND RANGE ON CARTESIAN GRAPHS

Let's transfer this knowledge to Cartesian graphs.

5. Page 5 shows a rounding function. Drag the input marker and think about how you can tell domain and range on a Cartesian graph.

6. On page 6, explore the function $f(x) = 2x$. Then edit the function and explore the functions listed here, again thinking about how to tell domain and range on a Cartesian graph.

To edit the function, double-click its equation and then enter the new expression.

$$g(x) = \frac{x}{2} \qquad h(x) = x^2 \qquad j(x) = 2x + 1 \qquad k(x) = 2\sin\left(\frac{\pi x}{2}\right)$$

Q6 How can you tell the range of a function just by looking at its Cartesian graph? How about its domain?

EXPLORE MORE

Q7 On page 1, change f's equation so that its range is all even numbers. Then change it again to make it all odd numbers. Record the equations you used.

Q8 On page 4, change v so that both its domain and its range include only numbers less than or equal to 0. Change w so that its domain is all real numbers except 0 and its range is all real numbers except 2. Record the equations you used.

The motivation for developing and using dynagraphs comes from the often-noted difficulty students have in seeing the graphs of functions as dynamic representations of functional relationships between two quantities and not just as static pictures. By decoupling the input and output axes, and having a segment connect points on parallel axes, students are better able to see the input-output machine view of functions expressed graphically. Being able to drag the input marker gives students the further advantage of actually varying the independent variable and seeing the function as a *dynamic* relationship between input and output.

Dynagraphs can serve as a bridge between the input-output machine model with which students are often introduced to functions and function graphs in the Cartesian plane.

SKETCH AND INVESTIGATE

Q1 Answers will vary, but should basically describe functions as consistent input-output machines. In other words, they are relations or mappings between input values and output values such that any valid input value maps to a single output value.

Q2 The dynagraphs do represent functions because they map input values to output values and they are consistent—a particular input value will always point to the same output value.

Q3 Answers will vary, but should not involve numbers or formulas. Good answers will in general include dynamic descriptions ("As the input is dragged steadily from left to right, the output . . .") and note any symmetries present.

NUMBERS, NUMBERS, NUMBERS

Q4 See Q3.

Q5
a. $t(1) = -1$
b. $t(5) = 7$
c. $x = -1$
d. $g = -6$
e. $p = -6$
f. $m = \ldots, -7, -3, 1, 5, \ldots$
g. $v(4) = 2$
h. $v(-4)$ is undefined
i. $r = 9$
j. $z = 1$
k. $s = 3$
l. $a = 2.5$

EXPLORE MORE

Q6 Function j has an absolute minimum of 0 at 0. Function u has an absolute maximum of 6 at $(\ldots, -7, -3, 1, 5, \ldots)$, and has an absolute minimum of -6 at $(\ldots, -5, -1, 3, 7, \ldots)$. Function v has an absolute minimum of 0 at 0.

WHOLE-CLASS PRESENTATION

Use **Introducing Dynagraphs Present.gsp** to explore this highly dynamic visual representation of functions with your students. Dynagraphs differ from Cartesian graphs in that you can make the variables really vary, so emphasize the variation in the presentation by using the Animation buttons and leaving variables moving on the screen.

This whole-class presentation allows students to gain a dynamic perspective on the notion of function and emphasizes the way in which the variables really vary.

SKETCH AND INVESTIGATE

Q1 Begin by asking students to describe a function in their own words. Get responses from several students, and encourage a diversity of descriptions. Consider forming small groups of two or three students and asking each group to create its own written description, suitable for explaining functions to someone who isn't familiar with them.

1. Open **Introducing Dynagraphs Present.gsp.** Four dynagraphs appear, each in a different color.

2. Explain that the input and output markers represent the variables and that this model allows you to vary the variables by dragging the markers. Use the **Arrow** tool to drag input marker *a*. After dragging it a bit, use the *Animate a* button to leave it in motion.

Q2 Ask students whether the behavior they observe represents a function, based on their description of what a function is. Solicit different explanations from as many students as possible.

Q3 Ask students to describe the behavior of this first function. They will want to call the tick mark "zero" or "the origin," and they will want to describe movement to the left or right as "increasing" or "decreasing." These characterizations are based on numbers; resist them, and instead encourage students to describe the behavior in terms of position, movement, and symmetry. Consider asking students whether the function has a "fixed point"—a state in which the input marker and the output marker are at exactly the same position.

> Leave each function in motion while you drag and discuss the remaining functions.

Q4 Drag the input marker for the second function and have students observe. Leave it in motion while students describe the behavior. (Students often want to call this function a "constant function." Rather than describing this answer as wrong, ask them whether it's the output that is constant or whether there is something else about this function that they view as being "constant.")

Q5 Use the *Animate c* button to put the third function's input marker into motion. (Students will often laugh at this function, and you may want to ask them how often they have laughed at a mathematical function.) Have them describe this function in detail. They may want to give it a name.

3. The remaining pages show dynagraphs with numbers added to the axes. Have students answer the questions on each page, and then allow them to see the algebraic formulas underlying the behavior of the dynagraphs.

Introducing Dynagraphs

A *function* is a mechanism that gives you one specific output value for any value that you put in.

How many ways are there to represent a function? You've probably encountered various representations of functions, using tables, graphs, or equations. In this activity you'll explore a new way of representing functions: *dynagraphs*.

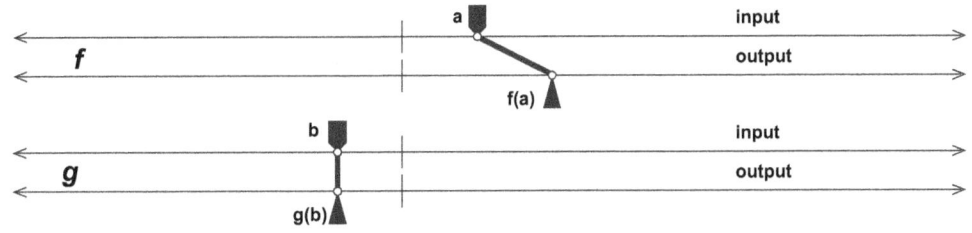

SKETCH AND INVESTIGATE

Q1 Spend a few moments reviewing with your group or on your own what a function is. Based on what you already know, how would you describe functions to someone who isn't familiar with them?

Each dynagraph has an input axis with an input marker and an output axis with an output marker. There is also a tick mark in the middle of each axis.

1. Open **Introducing Dynagraphs.gsp.** You'll see four dynagraphs labeled *f, g, h,* and *j,* each in a different color.

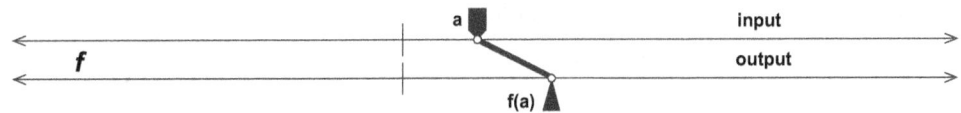

2. The input marker for dynagraph *f* is labeled *a.* To get an idea of how dynagraphs work, use the **Arrow** tool to drag this input marker.

Q2 Based on your understanding of functions, does this dynagraph represent a function? Explain.

Next you'll explore and describe in detail each of the dynagraphs on this page.

Here's a description of the *f* dynagraph:

This description is in terms of the position and motion of the input and output markers. The description does not use numbers or formulas because there are no numbers or formulas on the dynagraphs.

When the input marker is at the tick mark, the output marker is also at the tick mark. When the input marker is not at the tick mark, the output is always on the same side of the tick mark as the input. The output is always farther away from the tick mark than the input; it seems to be about twice as far away. When the input is dragged steadily from left to right, the output also moves steadily in the same direction, only faster.

Q3 Drag the input markers for dynagraphs *g, h,* and *j,* and then write detailed descriptions of these functions. Imagine you're describing the dynagraphs to someone who can't see them.

NUMBERS, NUMBERS, NUMBERS

You may have thought that it would be convenient to have number lines as the dynagraph axes to make it easier to give a precise description of the behavior of each function. With only a single tick mark, it's impossible to assign numbers to positions, such as "an input of 3 gives an output of 5."

In this section you'll explore four new dynagraphs, first without numbers, then with.

3. On page 2 are the new dynagraphs. Explore each by dragging its input marker.

Q4 Write a description of each function, just as in the previous section.

4. Press the *Show Number Lines* button. The dynagraph axes appear as number lines.

5. Drag *t*'s input control to 4.

The arrow points to an output of 5, as shown here. Using function notation, you can write $t(4) = 5$, which is read "*t* of four equals five."

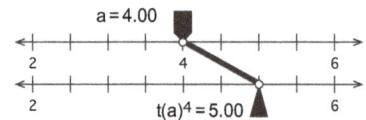

$a = 4.00$

$t(a)^4 = 5.00$

Q5 Solve for these unknowns. Be sure to use the correct function dynagraph for each one. Write each answer using function notation.

Hint: The answers to all but two of these questions are single numbers. One answer is "undefined," and one consists of several numbers.

a. $t(1) =$

b. $t(5) =$

c. $t(x) = -5, x =$

d. $u(-1) = g, g =$

e. $u(3) = p, p =$

f. $u(m) = 6, m =$

g. $v(4) =$

h. $v(-4) =$

i. $v(r) = 3, r =$

j. $w(2) = z, z =$

k. $w(4) = s, s =$

l. $w(a) = 0, a =$

EXPLORE MORE

A function has an *absolute maximum* if there is a largest output value—one the function can reach but can never exceed. Similarly, a function has an *absolute minimum* if there is a smallest output value—one the function can reach but can never go below.

Q6 Of the eight functions in the sketch, which functions have an absolute maximum or an absolute minimum? What are these maximum/minimum output values, and for what input values do they occur?

Relations and Functions

EXPLORE

Use this activity when introducing students to the concepts of relations and functions. The example of the rabbit is memorable and will remind students of their conclusion that time cannot be a function of location.

Q1 It is not possible to be in two places at the same time. It is possible to be in the same place at two different times. Explanations will vary, but this can generate an interesting discussion. Teachers generally describe the independent variable as a variable that you have control over—one that you have the ability to vary. But for these measurements, the dependent variable (location) is the one you can control, whereas the independent variable (time) proceeds rudely onward, oblivious to any attempt to control it.

Q2 Flopsy cannot be in two locations at the same time, but she is at position 9 (for instance) at both $t = 1$ and $t = 7$. Thus she's at the same place at two different times.

Q3 Both arrows lead away from 4 in the position bubble, leading to 2 and to 6 in the time bubble. Thus Flopsy was at this position at two different times: two seconds after starting and six seconds after starting.

Q4 Several values of position correspond to more than one ordered pair (in other words, to more than one row of the table). There are two pairs with a first element of 16, two with a first element of 9, two with a first element of 4, and two with a first element of 1.

Q5 When $t = 2$, Flopsy was at position 4.

Q6 There are no ordered pairs with the same time and different locations—that is, none with the same first element and a different second element. This is the fundamental definition of a function.

Q7 The relation (t, s) on page 3 is a function. The relation (s, t) on page 2 is not. You can tell from the arrows because there's only a single arrow leading from any given first element.

EXPLORE MORE

Q8 You can drag the points so that they have the same first element and different second elements. For instance, the ordered pairs (3.50, 1.55) and (3.50, 6.72) are both part of the relation. Therefore, this relation is not a function. These points are aligned with each other vertically.

Q9 Problem b is a function; there are no ordered pairs with the same first element and different second elements. Problem c is also a function. Problem d is not a function; numeric data to demonstrate this will vary, but students should observe that they are able to arrange the two points so they are aligned vertically.

Q10 This is not a function, because there are several places where the same first element corresponds to more than one second element. The vertical line demonstrates this condition by intersecting the function plot in two or more places.

Q11 Problem b is a function, and the vertical line can never intersect it at more than a single place. Problem c is the inverse of the cosine function; this inverse is not a function itself, because the vertical line intersects it in many places. The inverse cosine must be defined to have a restricted range if it is to be dealt with as a function. Problem d is a step function. Though the vertical line comes close to two different segments of the graph at certain positions, in fact it never intersects the two segments at the same time.

The presentation should follow the sequence in the student activity. Begin with an informal description of relations and functions, and use the presentation to motivate a more precise definition.

Start with these descriptions: A *mathematical relation* exists when two mathematical values are related in some way. A *function* is a specific kind of relation.

EXPLORE

Q1 Can you be in two different places at the same time? Can you be in the same place at two different times? Ask a number of students to respond and explain.

1. Open **Relations Functions Present.gsp** and press *Run!* to see Flopsy steal a carrot. Run the animation again and point out the variables *s* (which shows Flopsy's position) and *t* (which shows the time).

2. Reset Flopsy. To record her position and time, press *Show Table* and double-click the table to make the first row of values permanent.

3. Press *Dart!* repeatedly to make Flopsy dart forward. At the end of each movement, double-click the table to record the new position and time. You should have nine rows of permanent data when you finish.

> This table does not define a function. Our purpose is to clarify the distinction, so the first relation students work with should not be a function.

Give students a more precise definition, now that it's illustrated by a table: A *relation* is any set of ordered pairs, such as the ordered pairs (*s*, *t*) that appear in your table.

6. On page 2, press the *When was she at s = 4?* button. Question students, and discuss the meaning of the arrows, until they are all convinced that there's not just one right answer to this question.

7. These two arrows show part of the mapping from position (*s*) to time (*t*). Reset the maps and then use either Show button to show the full mapping.

> For a relation to be a *function*, each input value must correspond to only a single output value.

8. Contrast the full mappings (with all arrows showing) from pages 2 and 3 to further clarify the distinction between the relation that is a function (page 3) and the one that is not (page 2).

9. Use page 4 to see which problems allow ordered pairs that violate the function definition and to point out that the two ordered pairs are aligned vertically when they prove a relation is not a function.

10. Use page 5 to demonstrate the vertical line test for functions.

Finish with a class discussion reviewing the definitions. Ask students why they think functions might be so important in mathematics. [One benefit of working with functions is that the ability to find a unique output, given any input, makes many problems much easier to manage.]

Relations and Functions

A *mathematical relation* exists when two mathematical values are related in some way. A *function* is a specific kind of relation. In this activity you'll explore several relations, decide which of them are functions, and develop a graphical test to determine whether a relation is a function.

EXPLORE

Q1 Think about the mathematical relation between your location and the time. Is it possible for you to be in two different places at the same time? Is it possible for you to be in the same place at two different times? Explain your answers.

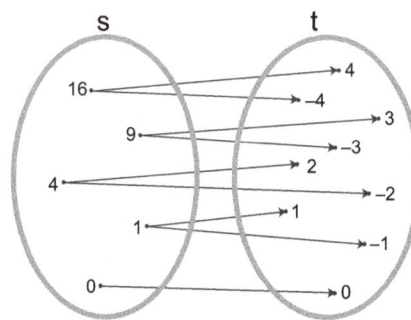

Understanding the answers to these two questions will help you to differentiate between relations that are functions and those that are not.

1. To explore a relation between position and time, open **Relations and Functions.gsp.**

2. Press *Run!* to see Flopsy steal a carrot. Notice the variables *s* (which shows Flopsy's position) and *t* (which shows the time). When Flopsy finishes running, press *Reset.*

3. Create a table to record the position and time. Double-click the table to make the first row of values permanent.

4. Press *Dart!* to make Flopsy dart forward. Double-click the table to record her new position and time.

5. Continue moving Flopsy (by pressing *Dart!*) and recording the data (by double-clicking the table) until Flopsy has returned to her original position. The table should contain nine rows of permanent data when Flopsy finishes.

> To create the table, select measurements *s* and *t* in order, and choose **Number | Tabulate.**

Definition: A *relation* is any set of ordered pairs, such as the ordered pairs (*s*, *t*) that appear in your table.

Q2 Review the data in this relation involving *s* and *t*. Is it possible for Flopsy to be in two different places at the same time? Is it possible for her to be in the same place at two different times? Explain by referring to the data.

6. Page 2 of the sketch shows the same table for *s* and *t*, and also shows two large bubbles containing all the numbers that appear in the table.

Q3 Press the *When was she at s = 4?* button. Answer this question using the arrows that appear. How does this relate to the questions that opened this activity?

7. These two arrows show part of the mapping from position (*s*) to time (*t*). Reset the maps and then use either Show button to show the full mapping.

Q4 From the arrows, are there any ordered pairs that have the same first element (*s*) but different second elements (*t*)? Which ones?

8. Go to page 3. You will look at the same data, in the form (*t, s*) rather than (*s, t*).

Q5 Press the *Where was she at t = 2?* button. Where was Flopsy when the time measurement was 2?

9. The arrows on this page show the mapping from time (*t*) to position (*s*). Reset the maps and then use either Show button to show the full mapping.

Q6 From the arrows, are there any ordered pairs that have the same first element (*t*), but different second elements (*s*)? Which ones?

When a relation is
a function, the first
element is called the
independent variable,
and the second is the
dependent variable.

Definition: A *function* is a relation for which there is exactly one second element for each first element. (If *t* is the first element and *s* is the second element, this means that Flopsy cannot be in two different places at the same time.)

Q7 Is the relation (*t, s*) shown on page 3 a function? What about the relation (*s, t*) shown on page 2? How can you tell from the arrows?

EXPLORE MORE

You will now develop a visual test for determining whether a relation is a function.

Q8 Page 4 contains two movable points that are part of a relation. Try to drag the two points so that they both have the same first element but different second elements. Can you do this? Is the relation a function? If not, what values did you use to prove it's not? Where are the points located relative to each other?

Q9 Press the *Problem b* button and try dragging these points. Can you make both of them have the same first element but different second elements? If so, what are the values, and where are the points relative to each other? Is it a function? Also try problems c and d on this page.

Q10 Page 5 shows the graph of a relation, and also shows a movable vertical line. Drag the line back and forth to be sure it remains vertical. How can you use the vertical line to tell whether the graph represents a function?

Q11 Investigate relations b, c, and d on page 5. Which are functions? For each one that's not, at what coordinate did you place the vertical line to prove that it's not?

Trigonometry and Statistics (Algebra II and Trigonometry)

Right Triangle Functions

RATIOS

Q1 $\sin 30° = \frac{1}{2}$. If students drag point B to make the angle 30°, the value could be slightly different. By using the button, they will get the exact value.

Q2 Answers will vary. The important thing is that students make predictions before trying it.

Q3 As students change the slider, the size of the triangle changes, but the angle and ratio remain constant.

Q4 The angles for the triangle remain constant as the size of the triangle changes. The resized triangle is similar to the original, so there is a scale factor by which each side has been multiplied to generate the new triangle. Therefore, each ratio must remain constant, because its numerator and denominator have been multiplied by the same factor.

ANGLES UP TO 90°

Q5 Students should end up with a sketch of the sine function between 0° and 90°.

12. It's important that students sketch the graphs themselves on paper, rather than printing out the sketch. The physical act of sketching the shapes helps them to remember the characteristics of the graphs.

Q6 The value of the sine increases from 0° to 90°, quickly at first and then more and more slowly. The value of the cosine decreases from 0° to 90°, slowly at first and then more quickly. The value of the tangent increases from 0° to 90°, very quickly as it approaches 90°.

Q7 No, $\cos 90° \neq \cos 30° + \cos 60°$. A reason can be found by looking at the plot for cosine. Using the plot, you can see that the value of $\cos 30° + \cos 60°$ is definitely greater than the value of $\cos 90°$. In general, there are very few functions where $f(x+y) = f(x) + f(y)$.

Q8 The values of $\cos 60°$ and $\sin 30°$ can be represented in one triangle. If $\angle A$ is 60°, $\angle B$ must be 30° since this is a right triangle. This illustrates that cofunctions of complementary angles are equal, because they refer to the same parts of the triangle.

Q9 The tangent value is undefined. The length of the adjacent side is 0, and division by zero is undefined.

EXPLORE MORE

Q10 If the triangle had an obtuse angle and a right angle, the sum of the angles would be more than 180°.

Q11 When you move B past the vertical to try to make an angle of 150°, you end up with a right triangle with an angle of 30°. (This angle is called the *reference angle* for 150°.) You could use this result to define $\sin 150° = \sin 30°$. (Although this result is correct, $\cos 150° \neq \cos 30°$, because the signs are different. Proper definitions of these functions for angles over 90° is best done by using the unit circle.)

Q12 The Calculator results indicate that $\sin 150° = \sin 30°$ and that $\sin 210° = -\sin 30°$. Students may speculate that the difference is that the opposite side goes up for 150° but down for 210°.

Q13 Another angle with the same sine is 330°.

WHOLE-CLASS PRESENTATION

Use the Presenter Notes and **Right Triangle Functions Present.gsp** to present this activity to the whole class.

Right Triangle Functions

Use this presentation to review the right triangle definitions of the trigonometric functions and extend them to angles greater than 90°.

1. Open **Right Triangle Functions Present.gsp.** The measure of $\angle CAB$ is on the screen. Drag point B to show its range.

Q1 For $\angle CAB$, identify the opposite side, adjacent side, and hypotenuse.

2. Press *Show Labels* and *Show Lengths.* Review the sine, cosine, and tangent definitions.

Q2 Set $\angle CAB$ to an arbitrary angle. Which of the three ratios is the greatest? Which is the least? [The answer depends on the angle. Students should be able to make a good guess by looking at the sides.] Press *Show Ratios* to reveal the answers. Repeat this part.

3. Pick another arbitrary angle. Have a student write the three ratios on the board.

Q3 What is the complement of this angle? What are the sine, cosine, and tangent of the complement?

4. Press the *Complement* button and discuss the relationships between the ratios for the original angle and the ratios for the complement.

5. Go to page 2. Again, you can change the angle by dragging a point.

6. Press *sin x* to see the point $(x, \sin x)$. Change the angle to create a trace of the sine graph. Do the same for the cosine and the tangent.

Q4 What is the geometric relationship between the sine and cosine graphs, and how is this explained by the relationship $\sin x = \cos(90° - x)$?

Q5 Why is there no upper limit to the tangent function?

Page 3 introduces trigonometric functions for angles greater than 90°. Discuss the fact that although an angle can be greater than 90°, an angle of that size will not fit into a right triangle. In that case, there is a related acute angle that serves as a *reference angle.* On page 3, $\angle CAB$ is the reference angle for $\angle DAB$.

7. Press *Show Angle CAB* and *Show Angle DAB.* Drag point B around to change the angles.

Q6 How are the sine, cosine, and tangent of these angles related? Students will see that the functions of $\angle DAB$ are sometimes negative, but always have the same magnitude as the corresponding functions of $\angle CAB$. Keep dragging point B and guide the class in determining the domain in which each function is negative.

8. On page 4 trace the function graphs again, this time for all angles between 0° and 360°.

Right Triangle Functions

For GSP5

Within a right triangle there are many important and useful ratios. You may have used *SOH CAH TOA* as a way of remembering the names of certain right triangle ratios:

> *SOH*: The ratio for *Sine* is *Opposite* over *Hypotenuse*.
>
> *CAH*: The ratio for *Cosine* is *Adjacent* over *Hypotenuse*.
>
> *TOA*: The ratio for *Tangent* is *Opposite* over *Adjacent*.

In this activity you'll explore these ratios in triangles of different shapes and sizes.

RATIOS

1. Open **Right Triangle Functions.gsp.** Drag each point and observe its effect on the size and shape of the triangle.

2. Use the **Text** tool to label the sides of the triangle *Opposite*, *Adjacent*, and *Hypotenuse* based on their position relative to ∠*CAB*.

$m\angle CAB = 30.00°$
Opposite = 2.52 cm
Adjacent = 4.37 cm
Hypotenuse = 5.04 cm

3. Measure ∠*CAB* and the length of each side.

*To calculate the ratio, choose **Number | Calculate** and enter each side into the calculation by clicking its measurement in the sketch.*

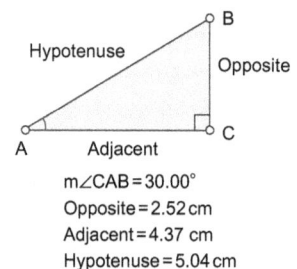

4. Calculate the sine ratio for ∠*CAB* by dividing the appropriate sides of the triangle. Label the resulting ratio *sin A*.

5. Drag *B* and observe its effect on sin *A*. Use the button to make $m\angle CAB = 30°$.

*Double-click the **Text** tool on the resulting calculation to change its label.*

Q1 What is sin 30°?

Q2 If you make the triangle larger without changing $m\angle CAB$, do you think sin *A* will increase, decrease, or stay the same? Give a reason for your prediction.

Q3 Change the size of the triangle by dragging the *Length of Hypotenuse* slider. What happens to $m\angle CAB$? What happens to sin *A*?

6. Construct the other two ratios based on *SOH CAH TOA*. Label the ratios *cos A* and *tan A*.

What term is used for triangles of different size but the same shape?

Q4 Why do the trigonometric functions stay the same if the triangle is made larger or smaller without changing the angles?

ANGLES UP TO 90°

You have just determined how the triangle's size affects (or does not affect) the values of the ratios. Now you'll make a table and graph to investigate how the angle affects the ratios.

7. Place $m\angle CAB$, $\sin A$, $\cos A$, and $\tan A$ in a table by selecting all four values in order and choosing **Number | Tabulate.**

8. Change the angle and watch the numbers change. Double-click the table to make the first row permanent. Continue dragging B and double-clicking the table until you have ten rows of values for different angles between 0° and 90°.

9. Press *Show Axes*. Then select the table and choose **Graph | Plot Table Data.** Choose $m\angle CAB$ for x and $\sin A$ for y.

Q5 On your paper, sketch the shape of the plotted points.

> To plot a point, select two measurements in order and choose **Graph | Plot as (x, y).**
>
> To trace a point, select it and choose **Display | Trace Plotted Point.**

10. To get a more complete graph of the relationship, plot the point ($m\angle CAB$, $\sin A$). Turn on tracing for the plotted point.

11. Fill in the gaps of your graph by dragging B to change the angle from 0° to 90°.

12. Following the same steps, plot both ($m\angle CAB$, $\cos A$) and ($m\angle CAB$, $\tan A$). Sketch and label all three graphs on your paper.

Q6 Describe the behavior of each graph. Is it increasing, decreasing, or constant? For what angles is it changing quickly? For what angles is it changing slowly?

> You can use the buttons to set the angle exactly.

Q7 Does $\cos 90° = \cos 30° + \cos 60°$? Use your cosine graph to explain your answer.

> As a hint, set angle A to 60°. What is angle B? Look at the sides involved for $\cos A$ and $\sin B$.

Q8 Why does $\cos 60° = \sin 30°$? Use your answer to find a second pair of angles for which the cosine of one is equal to the sine of the other.

Q9 What happens to the tangent when the angle is 90°?

EXPLORE MORE

Since a right triangle cannot have an obtuse angle, can $\sin 150°$ exist? What would the triangle look like if you could drag B so that $m\angle CAB > 90°$? What would happen to the ratios?

Q10 Why can a right triangle not have an obtuse angle?

Q11 On page 2 point B is free to move in a complete circle. What happens when you drag the angle past 90°? How could you use the result to define $\sin 150°$?

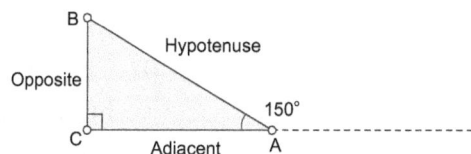

13. Using Sketchpad's Calculator, calculate $\sin 150°$ and $\sin 210°$.

Q12 How do these values compare to $\sin 30°$?

Q13 Find another angle greater than 180° whose sine is the same as $\sin 210°$.

Radian Measure

WHAT IS A RADIAN?

Q1 All three measurements (the radius of the circle, the length of the blue segment, and the length of arc *a*) are the same. This is because the blue segment started out as a radius of the circle, and when it rolled along the circle, it measured out an arc that is the same length as it is.

Q2 1 radian $\approx 57.30°$

Q3 A semicircle has a central angle of exactly π radians. If students are not aware of that fact, they should be able to give an estimate between 3.0 and 3.3. There are 2π radians in a complete circle.

Q4 Start with the circumference formula:

$$\text{circumference} = 2\pi r$$

This means that there are 2π radius lengths in the circumference, so the angular measure is 2π radians.

WHY RADIANS?

Q5 The measurements should verify this fundamental relationship:

$$\text{arc length} = \theta r$$

The formula will work for any angle in the given range, and for any radius. It will not work when the angle units are degrees.

Students may notice some conflicts with the units. The arc length will be in centimeters, but the θr calculation will be in radians · centimeters. Leave that discrepancy for the discussion.

Q6 When degrees are converted to radians, you get this simpler formula:

$$\text{area} = \frac{\theta}{360°}\pi r^2 = \frac{\theta}{2\pi}\pi r^2 = \frac{\theta r^2}{2}$$

Q7 The formula works for any angle between 0 and 2π, and for any radius, but it does not work when the angle units are degrees. Again, the number values agree, but there is an apparent discrepancy with the units.

DISCUSS

The discussion questions will be more helpful if the entire class works together. Here are some suggested points.

Q8 Although radian angle measurement is useful, an angle of one radian has no great significance. The really useful angles (180°, 90°, 60°) have measures that are irrational numbers when expressed in radians, so they cannot be expressed exactly with a decimal expansion. However, we can express them as simple fraction multiples of π (π, $\pi/2$, $\pi/3$).

Q9 A radian measurement is an angle measurement, but you can just as well think of it as the ratio of the lengths of an arc and its radius. Since it is a ratio of two linear measurements, it has no units. An advantage of this concept is that it clears up the unit discrepancy that appeared in Q5 and Q7.

Q10 The number of radians in a circle is 2π, an irrational number. It is impossible to divide a circle into an integral number of radians. It is also impossible to do this with a tenth, hundredth, thousandth, or any other fraction of a radian. Instruments that measure angles (protractors, compasses, theodolites, sextants) need to have divisions that are all the same, so the angle unit must divide the circle evenly.

One obvious solution might be simply to graduate the instruments in some fraction of π radians. In fact, that is exactly what we do ($1° = \pi/180$ radians).

1. Open **Radian Measure Present.gsp.** Press the *Go* button.

As the radius segment rolls around the circle, explain that students can think of a radian as the angle that corresponds to one length of the radius being laid out along the circumference of the circle.

Q1 How long is the blue segment? [It's equal to the radius of the circle.]

Q2 What do the red ticks mark off? [Each tick marks a distance of one radius and an angle of one radian.]

2. Press *Reset;* then *1 Radian.* The animation stops after marking off one radian.

Q3 About how many degrees are there in one radian? To coax a good guess, point out the triangle formed by points *A, B,* and *C.* You can think of arc *BC* as a side of the triangle. That would make it an equilateral triangle, but one of the sides is not straight. Would that make $\angle BAC$ greater than 60° or less than 60°?

3. Press *Show Central Angle.* It will show that $\theta \approx 57.30°$.

4. Press *Semicircle.*

Q4 Count the tick marks. How many radians are there in a semicircle? [a little more than 3]

To change the angle units, choose **Edit | Preferences.**

5. Change the Angle Units to **radians.** Angle θ appears as 1π radians.

6. So a semicircle has π radians. That means that a circle must have 2π radians. Press *1 Circle* to confirm that.

Q5 But you already knew that, didn't you? What is the circumference in terms of *r*? [$2\pi r$] So how many times will the radius go into the circumference? [2π] So how many radians are there in a circle? [2π]

7. Press *Reset;* then press *Go.* Press *Go* again to stop the animation with θ somewhere between 0 and 2π. Press *Show Arc.*

When entering numbers that are on the screen, click the measurement itself.

8. Here is something you can do with radians, but not with degrees. Choose **Number | Calculate.** Enter $\theta \cdot r$. Compare the calculation with the measured length of the arc. Try it with several different values of θ and *r*.

Q6 The formula for area of a sector is $\frac{\theta}{360°} \pi r^2$. Convert the 360° to radians and simplify. What is the new formula? $\left[\frac{\theta r^2}{2}\right]$

9. Press *Show Sector.* Use the Sketchpad Calculator and enter $\theta \cdot r^2/2$. Compare the answer to the measured arc length.

Radian Measure

So you think you know angle measurement? An understanding of degrees is a valuable skill, but there are other ways to measure. Other angle units include points, grads, mils, and dekans. To make matters worse, these units may go by different names in different places. Still worse, they may have the same name, but different definitions.

Among all of these angle units, the radian holds a special place. You can use it to measure angles, of course, but radian measure also describes relationships between certain geometric objects.

WHAT IS A RADIAN?

1. Open **Radian Measure.gsp.** Press the *Go* button and watch the circle radius rotate into a tangent position and then roll around the circle.

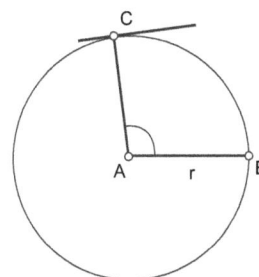

You will use this line segment to measure a central angle of the circle. This will be the basis for defining radian measure for angles.

2. Press *Reset* and then *Home* to stop the animation and return the radius to its tangent position. Measure the radius of the circle and the length of the blue segment.

3. Press the buttons *Show Central Angle, Show Arc,* and *1 Radian.* Measure the length of arc *a.*

Q1 The central angle, $\angle BAC$, is now exactly one radian. What do you notice about your three measurements? Explain why they come out this way.

Q2 The measure of the angle, θ, is displayed in degrees. Approximately how many degrees are there in one radian?

Notice that although θ is equivalent to $m\angle BAC$, it can keep increasing past 360°.

4. Press *Semicircle.* The line segment will continue to roll until it has stepped off half of the circle.

Q3 Using the tick marks to approximate an answer, how many radians are in a semicircle? How many radians will there be in a complete circle?

5. Press the *1 Circle* button to check your last answer.

6. Choose **Edit | Preferences.** Change the Angle Units to **radians.**

Q4 The angle measurement now shows you exactly how many radians are in a circle. But you already knew that, didn't you? Write the formula for the circumference in terms of the radius. Use that along with the definition of a radian to prove that there are exactly 2π radians in a circle.

WHY RADIANS?

So far, you have not seen any good reason for using radians rather than degrees. Actually, we use radians in order to make things easier, not harder.

Do these measurements one at a time. Select one object and choose the appropriate command from the Measure menu.

To change the radius, drag point *B*.

7. Press *Reset* and *Go*. Press *Go* again to stop the animation before the angle makes a complete circle ($0 < \theta < 2\pi$).

8. You have a measurement for angle θ and a measurement for radius *r*. Use the calculator to find the product $\theta \cdot r$.

Q5 What is the arc length in terms of θ and *r*? Check your answer with different radii and different angles in the range $0 < \theta < 2\pi$. Does your formula always work? Does it work when you use degrees?

The area of a circle sector varies directly with the central angle. You probably are familiar with this formula:

$$\text{sector area} = \frac{\theta}{360°}\pi r^2$$

Q6 Rewrite the above formula using radians instead of degrees. Simplify your answer.

9. Select the arc and choose **Construct | Arc Interior | Arc Sector.** Select the sector and choose **Measure | Area.**

Q7 Using your formula from Q6, calculate the area of the sector. Does it match your measurement in all cases?

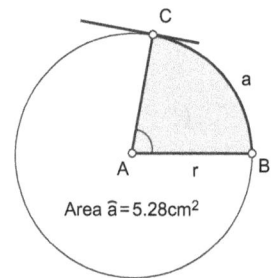

Area â = 5.28cm²

DISCUSS

Q8 When using radians, Sketchpad automatically expresses angle measurements in multiples of π. This is a common practice. Why?

Q9 It is also common practice (not used by Sketchpad) to write radian angle measurements without writing any units at all. Why is that?

Q10 In spite of the radian advantages you have seen here, degrees are more common in practical applications. What advantages do degrees have?

Unit Circle and Right Triangle Functions

THE UNIT CIRCLE

Q1 The value of *Arc Angle* ranges from 0° to 360°.

Q2 The sine ranges from −1 to 1, at angles of 270° and 90°, respectively. The cosine ranges from −1 to 1, at angles of 180° and 0°, respectively. The tangent has no limit in either direction, but the measurement is limited by the resolution of the objects on the screen.

THE REFERENCE TRIANGLE

Q3 The calculation corresponds to the sine function.

Q4 The smallest value for sine is 0 and occurs at 0°. The largest value is 1 and occurs at 90°. The smallest value for cosine is 0 and occurs at 90°. The largest value is 1 and occurs at 0°. The smallest value for the tangent occurs at 0°. The tangent has no upper limit, and it gets very large as the angle approaches 90°. At both 0° and 90°, the triangle is degenerate, with various points and sides coinciding.

COMPARE THE DEFINITIONS

Q5 The measurements agree only in the first quadrant. In the other quadrants the arc angle is more than 90°, but the angle in the triangle remains between 0° and 90°.

Q6 The definitions agree in Quadrants I and II because the *y*-value is positive there. The definitions disagree in the other two quadrants because the measured length of a line segment is always positive.

Q7 The cosine values agree in Quadrants I and IV, but disagree in Quadrants II and III. In these two quadrants the *x*-value is negative, but the distance measured in the triangle remains positive.

Q8 The tangent values agree in Quadrants I and III. In Quadrant I the coordinates (for the unit circle definition) and the distance measurements (for the right triangle definition) are all positive, so the two functions agree. In Quadrant III both coordinates are negative, so their ratio is positive, matching the right triangle definition. In the other two quadrants one coordinate or the other is negative, resulting in values that the right triangle cannot produce.

Q9 Explanations will vary. This is a good place to introduce the idea of the *reference triangle* within the unit circle and to observe that the opposite side for both 30° and 150° corresponds to the same *y*-value. For 210°, the opposite side corresponds to a negative *y*-value, so the value of sin 210° is the opposite of that of sin 30°.

Q10 Answers will vary. A big advantage of the unit circle method is the ability to work with angles that are beyond 90°. An advantage of the right triangle method is that it's easier to apply when the angle is not in standard position. (Though students don't know this yet, the unit circle method will allow them to explore topics, such as uniform circular motion, which would not be possible with only a right triangle definition.)

EXPLORE MORE

Q11 Answers will vary. Analyzing the flight path of a plane or the position of a person on a Ferris wheel both benefit from using angles beyond 90°. For the height of a building, a right triangle definition is sufficient.

Q12 You could condense the two methods into one by considering the right triangle method to be a special case of the unit circle in Quadrant I.

Q13 At 90° the line *AC* is vertical, so its slope (and the tangent of 90°) is undefined. The result is that the tangent graph has an asymptote at 90°.

RELATED ACTIVITIES

The definitions are introduced in Right Triangle Functions and in Unit Circle Functions.

Use this presentation to relate the two ways of defining the trig functions.

THE UNIT CIRCLE

Angles in this sketch are in degrees rather than radians.

Leave point *C* in motion while students answer these three questions.

1. Open **Unit Circle Right Triangle Present.gsp.** Show the unit circle and animate point *C*. Show the arc angle.

Arc Angle = 46.68°

Q1 Ask students what are the largest and smallest values they observe for the arc angle. (0° to 360°)

Q2 Ask what measurements are needed in the unit circle to define the sin, cos, and tan functions. (in order: *y*, *x*, and equivalently either *y*/*x* or the slope of *AC*)

Q3 Show these measurements and ask students to observe the largest and smallest values for each of the measurements. Review again which is sin, cos, and tan.

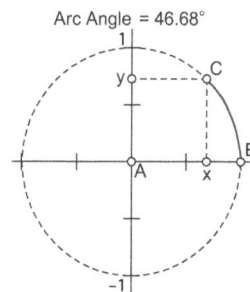

THE REFERENCE TRIANGLE

2. Show the right triangle and measure ∠*DEF*.

$m\angle DEF = 30.87°$

Hypotenuse   Opposite   Adjacent

Q4 Drag point *F* and ask students to observe the largest and smallest values for the angle.

Q5 Show the length measurements and ask students what ratios must be calculated to find the sine, cosine, and tangent.

Q6 Show the ratios and have students confirm which is sine, cosine, and tangent. Drag point *F* and have students observe the largest and smallest values for each ratio.

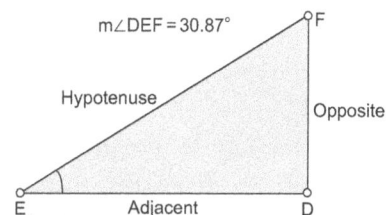

COMPARE THE DEFINITIONS

3. To compare the definitions, combine the models. Press *Merge Triangle to Circle.*

Q7 Drag point *C* (keeping it in Quadrant I), and ask students to compare the four measurements from each triangle.

Q8 Ask students to make conjectures about what will happen if *C* leaves Quadrant I.

Q9 Drag point *C* slowly through the other three quadrants, and ask students to describe what they observe about each of the four measurements. Encourage them to explain their observations.

Q10 Why does the sine of 150° in the circle have the same value as the sine of 30° in the triangle? Why is sin 210° the opposite of sin 30°?

Q11 Ask for advantages and disadvantages of each way of defining the functions.

Unit Circle and Right Triangle Functions

There are several different ways to define trigonometric functions like sine and cosine. One set of definitions is based on right triangles, and another set is based on a *unit circle* (a circle with a radius of exactly one unit). In this activity you'll explore the relationship between these two ways of defining trigonometric functions.

THE UNIT CIRCLE

Arc Angle = 46.68°

1. Open **Unit Circle Right Triangle.gsp.** Measure the arc angle of arc *BC* on the unit circle. Label the measurement *Arc Angle.*

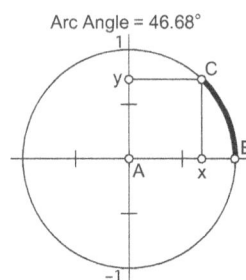

Q1 Drag point *C* around the circle and observe the angle measurement. What are the smallest and largest values that you observe? Leave *C* in Quadrant I when you finish.

2. Measure the *y*-coordinate of point *C* and label it *sin in circle.* Measure the *x*-coordinate and label it *cos in circle.*

3. Construct a line through *A* and *C,* and measure the slope of the line. Label this measurement *tan in circle.*

Q2 Drag *C* again. What are the smallest and largest values that you observe for the sine, cosine, and tangent of the arc angle? At what angles do these values occur?

THE REFERENCE TRIANGLE

Ratios of triangle sides provide another way to define trigonometric functions. You can use the mnemonic *SOH CAH TOA* to recall the ratios:

SOH: The ratio for *Sine* is *Opposite* over *Hypotenuse.*

CAH: The ratio for *Cosine* is *Adjacent* over *Hypotenuse.*

TOA: The ratio for *Tangent* is *Opposite* over *Adjacent.*

To measure ∠*E,* select points *D, E,* and *F.* Then choose **Measure | Angle.**

4. Measure ∠*E* for the right triangle.

5. Measure the *Adjacent* side by selecting points *D* and *E* and choosing **Measure | Coordinate Distance.** Do the same for the other two sides. Label your measurements *Opposite, Adjacent,* and *Hypotenuse.*

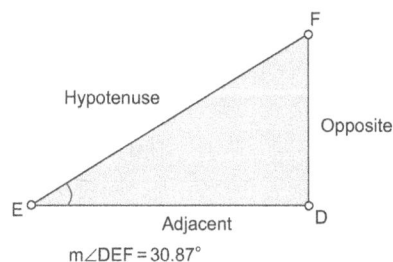

m∠DEF = 30.87°

6. Use Sketchpad's Calculator to calculate *Opposite/Hypotenuse.*

Q3 According to the *SOH CAH TOA* mnemonic, to which trigonometric function does this calculation correspond?

7. Label your calculation *sin in triangle*. Calculate each of the other two ratios and label them appropriately.

Q4 Drag point *F*. What are the smallest and largest values that you observe for the sine, cosine, and tangent in the right triangle? At what angle do the maximum and the minimum occur for each?

COMPARE THE DEFINITIONS

To compare these definitions, you'll combine the two models.

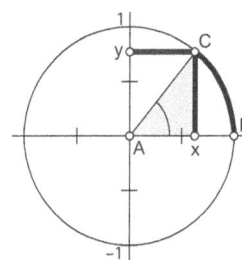

8. Select points *A* and *E*, and choose **Edit | Merge Points.** Also merge points *C* and *F*. The right triangle is now attached to the inside of the unit circle.

Q5 Drag point *C* and observe the two angle measurements (the arc angle and the angle in the triangle). When do these measurements agree? When do they disagree?

Q6 Drag point *C* and observe the two sine measurements. Explain why the values are equal in certain quadrants but not in others.

Q7 When do the two cosine measurements agree, and when do they disagree? Why?

Q8 When do the two tangent measurements agree? Explain.

Q9 Why is the sine of 150° the same value as the sine of 30°? Why is the sine of 210° the opposite of the sine of 30°? (*Hint:* Think about how each relates to either a coordinate or a ratio, and compare these.)

Q10 Describe possible advantages and disadvantages for each method of defining the trigonometric functions.

EXPLORE MORE

Q11 Based on the different definitions, which might be better to determine the flight path of an airplane? The position of a person on a Ferris wheel? The height of a building? Explain.

Q12 Could you always use a single definition? Explain.

Q13 Drag point *C*. What happens to the tangent at 90°? Explain. What does this mean in terms of the graph of the tangent function at 90°?

For GSP5 ACTIVITY NOTES

SKETCH AND INVESTIGATE

This activity uses the term *circular functions* because the functions are defined based on the position of *P* on the unit circle, rather than by reference to a particular triangle. If your students have never encountered this term, describe it as an alternate name for the trig functions. Explain that these are the same six functions, and that the choice of terminology is often based on the method used to define the functions.

6. The *sin* and *cos* segments correspond to the normal geometric definitions in △*OPS*, as shown in the figure on the left. The more common geometric definitions of the tangent and secant (segments *BC* and *OC* on the right) are replaced in this diagram by the congruent segments *PQ* and *OQ*, respectively.

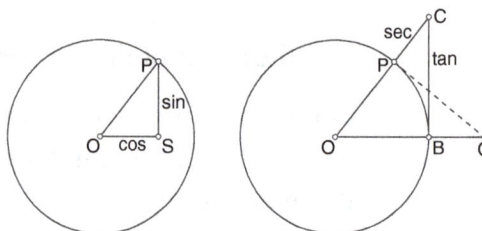

By using these alternate segments, the diagram becomes simpler and easier for students to remember. Another advantage of this arrangement is that the "co-" segments are all on the left, and the other segments (*sin, tan,* and *sec*) are on the right, reducing student confusion between sec and csc.

Q1–Q3 The completed table is shown here.

Segment	Label	QI	→	QII	→	QIII	→	QIV	→	QI
PS	sin	+	+	+	0	−	−1	−	0	+
PT	cos	+	0	−	−1	−	0	+	+	+
PQ	tan	+	∞	−	0	+	∞	−	0	+
PR	cot	+	0	−	∞	+	0	−	∞	+
OQ	sec	+	∞	−	−1	−	∞	+	+	+
OR	csc	+	+	+	∞	−	−1	−	∞	+

13. The value of ∠*BOP* ranges from −π to π. To create a graph from 0 to 2π, use the arc angle from *B* to *P* on the circle to measure the angle used for the plotted points.

16 and 17. In these steps, students establish the connection between the geometric segments and the shape of their graphs. This is an important connection that you should emphasize during class discussion of this activity. Once students have mastered this connection, they can re-create the shape of any of the six graphs simply by imagining the segment as point *P* moves around the circle.

DEMONSTRATE

To present the triangle and relationships in this sketch, use the All in One page of **Trigonometry Tracers.gsp.**

Six Circular Functions

For GSP5

Trig functions are often called *circular* functions when they are defined by measurements in a unit circle.

In this activity you will create a simple diagram that contains six segments corresponding to the six circular functions. You will use these segments to calculate the values of the functions and graph them.

SKETCH AND INVESTIGATE

1. In a new sketch, use the **Compass** tool to construct a circle.

2. Use the **Label** tool to label the center point *O* and the radius point *A*.

To define the coordinate system, select the circle and choose **Graph | Define Unit Circle.**

3. Define a new coordinate system using the circle as the unit circle. Hide the grid. Construct a point on the circle anywhere in Quadrant I and label it *P*. Construct the intersection of the circle with the positive *x*-axis, and label it *B*.

To construct the tangent line, select *P* and the radius segment and choose **Construct | Perpendicular Line.**

4. Construct the radius from *O* to *P*, and construct at *P* a tangent to the circle. Make both the radius and the tangent dashed. Label the tangent's intersection with the *x*-axis *Q* and the tangent's intersection with the *y*-axis *R*.

5. From *P* construct perpendiculars to both axes. Make both perpendiculars dashed. Label the intersection with the *x*-axis *S* and the intersection with the *y*-axis *T*.

6. Hide the tangent line and perpendiculars. Construct and label the six segments listed in the following table. Make each segment thick, and give each a different color.

7. For each segment, select its end points and measure the Coordinate Distance.

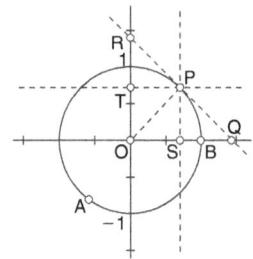

These distances correspond to the six circular functions, but because distance is always positive, you must pay attention to the behavior of the segment to determine when the corresponding function is positive and when it is negative.

Segment	Label	QI	→	QII	→	QIII	→	QIV	→	QI
PS	sin	+								+
PT	cos	+	0							+
PQ	tan	+	∞	−	0	+	∞	−	0	+
PR	cot	+								+
OQ	sec	+								+
OR	csc	+	1							+

Q1 Drag point *P* from Quadrant I to Quadrant II, and observe each distance at the transition. Some distances are 0, some are 1, and some increase without limit. Make a copy of the preceding table, and fill in the first "→" column with 0, 1,

310

New York City Title I High School Activities with The Geometer's Sketchpad
© 2012 Key Curriculum Press

or ∞ to indicate the behavior of the distance corresponding to each function. (Three of the cells in this column are already filled in for you.)

Q2 When a distance is 1 during the transition, the sign of the function remains the same, but when the distance is either 0 or unbounded, the sign of the function changes. In column QII, enter the new sign for each function in Quadrant II.

Q3 Observe the distances as you drag P from Quadrant II to Quadrant III. Fill in the next two columns of the table. Similarly, complete the rest of the table.

8. Measure the x- and y-coordinates of P. Use the Calculator to compute the values $\text{sgn}(x_P)$, $\text{sgn}(y_P)$, and $\text{sgn}(y_P/x_P)$.

> *Signum* (abbreviated sgn) is in the Calculator's Functions menu. This function returns a value based on the sign of the argument: 1 if it's positive, 0 if it's zero, or −1 if it's negative.

9. Observe the behavior of these three calculations as you drag P into each of the four quadrants. For each distance measurement, there's one calculation that produces the desired sign for the corresponding function in your table. For instance, the calculation $\text{sgn}(x_P)$ produces the desired sign for the cos function.

10. Multiply distance PT by the $\text{sgn}(x_P)$ calculation, and label the result *cos*. This calculation gives the correct value of the cosine function for every position of P.

11. Similarly, multiply each of the other distances by the calculation that will produce the correct positive and negative values according to your table. Label each result based on the corresponding circular function.

12. Color each calculation to match the color of the corresponding segment. Hide the intermediate calculations, the measurements, and the coordinate axes.

> To construct the arc, select the circle, point *B*, and point *P*, and choose **Construct | Arc on Circle.**

13. Construct the arc from B to P and measure its angle. If the measurement is in degrees, choose **Edit | Preferences** to set Angle Units to **radians.**

14. Construct a new point in empty space away from the unit circle, and choose **Graph | Define Origin** to define a new coordinate system. Hide the grid.

15. On the new coordinate system, plot the point $\left(m\widehat{BP}, \sin\right)$. Drag P to observe the behavior of the plotted point.

16. Construct the locus of the plotted point as P moves around the circle. Label this locus *sin*, and color it to match the corresponding segment and calculation.

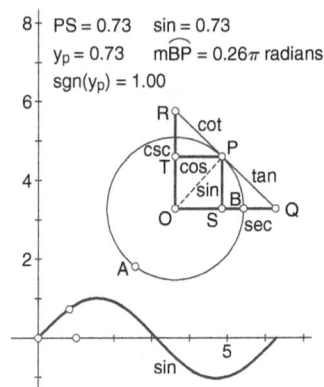

17. Similarly, plot points and construct loci to match the other five segments.

18. For each circular function, create a hide/show button to hide or show all of its features (the segment, the calculated value, and the locus). Create an animation button to animate P around the circle. Use these buttons to present your work.

Normal Distribution

RANDOM SIMULATION

(The test score is what's called a *binomial random variable,* and it results in a *binomial distribution.* Students don't need to be familiar with these terms to do the activity.)

Q1 The probability of scoring 60 or more is approximately 0.028. Don't expect students to calculate this; the question is only a prompt for discussion.

Q2 This question does not ask for a precise calculation either, but students should be able to give a refined answer based on the observations. Encourage them to keep track of results of new random samples.

The expected number of passing scores is about 6, which makes the probability about 0.03. Students may find this counterintuitive. Although scoring 6 out of 10 is no great feat, scoring 60 out of 100 is highly unlikely.

Q3 As *n* increases, the data distribution tends to fit the curve more closely.

Q4 To approximate the probability, calculate the ratio of the areas:

$$\text{probability} \approx \frac{\text{success area}}{\text{total area}} \approx 0.03$$

This method uses a continuous integration although the random variable is discrete. The regions represent not only the integers between 0 and 100 (the only possible scores), but also all real numbers between the integers and even a very small area outside of the range. This may cause some confusion. Remind students that it is only an approximation. There is some justification for changing the minimum score to 59.5, and that would make a slight difference in the result.

THE NORMAL DENSITY CURVE

Q5 Changing μ causes a horizontal translation of the curve, and μ corresponds to the score at the maximum point of the curve. Increasing σ stretches the curve horizontally and compresses it vertically.

The curve is above the *x*-axis for any real *x*. That can be confirmed from the function definition. Students may be intimidated by the complexity of the definition. Help them break it into manageable parts. Since σ is a positive number, the denominator must be positive. The numerator is a positive base, *e,* with an exponent, so it must be greater than zero no matter what the exponent is.

Q6 No matter what the settings are for parameters μ and σ, the total area under the curve is 1. That corresponds to the denominator of the area ratio from Q4. Therefore, the probability is simply the area of the success region.

Q7 This is the same probability again, 0.03. Students should set limit a to 59.5 and drag b far to the right. In theory there should be no upper limit, but that is not possible here, and the missing area is negligible.

This may raise some questions about σ. Here is the formula:

$$\sigma = \sqrt{kp(1-p)}$$

where p is the probability of success on any one problem, and k is the number of problems on the test.

Q8 You would be more likely to pass. Increasing σ flattens out the curve, forcing more area into the success region. The logical place to set the lower limit is 59 (since only even scores are possible), and the probability of passing is now about 0.10.

EXPLORE MORE

Q9 Based on the data, the probability is about 0.08. Students should set the parameters $\mu = 79.4$, $\sigma = 4.03$, $a = 70$, and $b = 74$.

Normal Distribution

Although statistical analysis uses mathematics, it is a science in that it involves the analysis of observations. Recent computer advances have made statistical analysis much easier and more efficient. Statistics is now an indispensable tool in such diverse fields as medicine, biology, economics, sociology, meteorology, and sports.

RANDOM SIMULATION

Suppose you are taking a test, but you are entirely unprepared. In fact, you have no knowledge of the subject matter at all. It's a true/false test with 100 questions, and you need to get at least 60 right to pass. You know that there are 200 students taking that same test, and not one of them knows anything about the subject.

Q1 None of this worries you. Anyone can answer half of the questions correctly simply by guessing, so you only need to be a little bit luckier than most. What would you guess is your probability of getting a passing score? Express your guess as a decimal number between 0 and 1.

1. Open page 1 of **Normal Distribution.gsp.** This sketch will display a random distribution of scores on the test. You can control three parameters:

   p, the probability of a correct answer on any one problem

   pass mark, the minimum score needed to pass the test

   n, the number of people taking the test

2. Edit these three parameters to model the true/false test described above. To try a new random sampling, select the red circle interior and press the exclamation point (!) key. Do this several times, and as you do so, watch the number of passing scores.

Double-click on a parameter to edit it.

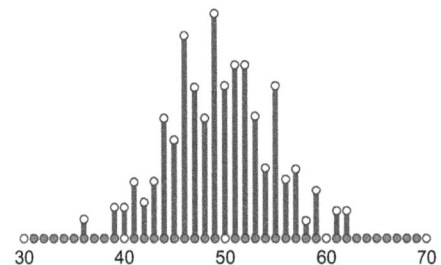

Q2 Based on these observations, about how many people out of 200 are likely to pass the test? What is the approximate probability of passing?

3. Select the red circle interior and press the exclamation point (!) key again several times. This time watch the overall shape of the distribution.

This is something you may have noticed in other data distributions. It has a bell shape, high in the middle and tapering down toward the extremes. Early investigators of probability theory noticed it too. They were successful in deriving a function that would predict the distribution.

4. Open page 2. The curve you see is a prediction of the data distribution. The vertical scale in this model is automatically adjusted so that the curve stays the same when you add more data.

The curve is actually growing taller when you increase n. The variable scale makes it appear stable and keeps it from outgrowing the screen.

5. Select the parameter *n* and press the **+** key several times. This will increase *n* by 100 each time. Run the number up to at least 1000.

Q3 What do you notice about the relationship between the random data and the curve as you increase *n*?

For large values of *n*, this function can give you a reliable approximation of the data distribution before you even see the data.

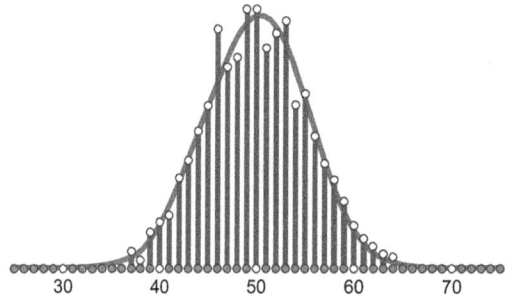

6. Press *Hide Data, Show Total Region,* and *Show Success Region.* When the regions appear, you will also see measurements representing their areas.

Q4 Based on the way you have seen the data fit the graph, how can you use the two area measurements to approximate the probability of passing the test? Do so now. What is the approximate probability?

THE NORMAL DENSITY CURVE

If you have a reasonable level of confidence in the shape of the distribution, you can estimate the probability of an event by comparing areas, just as you did above. The *normal density curve* is actually a family of curves. It has two parameters: μ (mu), the mean; and σ (sigma), the standard deviation. Here is the function definition:

Perhaps you are not yet familiar with e. For now, you only need to know that it is a constant. Its value is about 2.72.

$$f(x) = \frac{e^{-((x-\mu)^2/2\sigma^2)}}{\sqrt{2\pi}\cdot\sigma}, \sigma > 0$$

7. Open page 3. This is a normal density curve. Experiment with changing parameters μ and σ.

Q5 Describe in detail how changing μ and σ affects the shape of the curve. Does it cross the *x*-axis?

To set the limits more precisely, edit the parameters set a and set b, and then press Set Limits a and b.

8. Press the *Show Limits* button. The points *a* and *b* on the *x*-axis control the limits of a region under the curve. Notice the measurement for the area of this region. This time it has no units because it is based on a coordinate system.

Q6 With the earlier curve the height was based on the size of the data sample. That's not the case here. Drag the region limits to the edge of the screen left and right. What is the approximate total area under the curve? Try this again for several μ and σ settings. How does this simplify the probability calculation?

For the upper end of this interval, simply drag *b* to the edge of the screen. The missing area is negligible.

Q7 The true/false test example has normal parameters $\mu = 50$ and $\sigma = 5$. Enter these parameters and adjust the limits so that the region covers $x > 59.5$. What is your probability estimate?

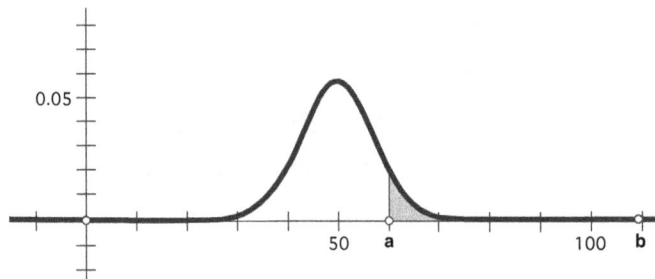

Q8 If the test contained 50 questions worth 2 points each, the parameters would be $\mu = 50$ and $\sigma \approx 7.07$. In that case, would you be more likely or less likely to pass the test?

EXPLORE MORE

It was possible to calculate μ and σ for these true/false test examples, but random variables are usually much more complex. Take a person's weight as an example. It is influenced by age, diet, health, and a combination of genetic material from thousands of generations of ancestors. No one could possibly gather and process so much information. What you have to do instead is to make measurements of the population or, more likely, a sample of the population. You can then derive μ and σ from that.

Sports is an excellent field for investigating statistics because there is plenty of readily available data on athletes' heights, weights, ages, performances, and even salaries. You will probably have to do your own calculations of the mean and standard deviation.

Q9 The published rosters of six National Basketball Association teams indicate a mean height of 79.4 inches, with a standard deviation of 4.03 inches. Based on this information, what is the probability of a randomly chosen NBA player having a height between 70 and 74 inches?

Permutation and Combination

ACTIVITY NOTES

PERMUTATION

Q1 There are six ways to arrange the prizes.

Q2 By adding one more prize, there are now 24 ways: For each prize chosen first, there are 6 ways to arrange the remaining three prizes.

Q3 For eight objects the number is $8! = 40,320$.

Q4 We usually say that $0! = 1$ by definition, for convenience, but there is a perfectly logical way of justifying this. Imagine starting with an empty set. If you do nothing at all, then you have one arrangement, and there is no way to change that arrangement. Therefore, there is one way to arrange zero objects.

SUBSETS

Q5 There are four prizes available for the first prize. After choosing that, there are three available for the second. There are 12 permutations.

Q6 Here is a symbolic proof that $_nP_0 = 1$:

$$_nP_0 = \frac{n!}{(n - 0)!} = \frac{n!}{n!} = 1$$

Another explanation is that $_nP_0$ is the number of ways to draw zero objects from a set of n. There is one way to do this, which is to draw nothing at all.

Q7 Symbolically:

$$_nP_n = \frac{n!}{(n - n)!} = \frac{n!}{0!} = \frac{n!}{1} = \frac{n!}{1!} = \frac{n!}{[n - (n - 1)]!} = {_nP_{n-1}}$$

Imagine drawing n objects from a set of n. You start by drawing $n - 1$ objects. There are $_nP_{n-1}$ ways to do this. In each case there is only one object left to complete the group.

COMBINATION

Q8 There must be six because that is the number of ways there are to arrange three objects ($3! = 6$).

Q9 $_nC_0 = 1$ $_nC_1 = n$ $_nC_n = 1$

Q10 Symbolically:

$$_nC_r = \frac{n!}{r!(n - r)!} = \frac{n!}{[n - (n - r)]!(n - r)!} = {_nC_{n-r}}$$

When you draw r objects, you could think of that as separating the set into two groups, the r objects that you drew and the $n - r$ objects that you left behind. If $_nC_r$ is the number of ways to draw r objects, then it must also be the number of ways to leave $n - r$ objects behind.

CALCULATIONS

Q11 You are drawing five players from a set of ten, and their positions do matter, so use the permutation formula.

$$_{10}P_5 = 30{,}240$$

Q12 This is simply the number of ways to arrange nine objects.

$$9! = 362{,}880$$

Q13 Each player gets five cards from a set of 24 different cards, and the order of the cards does not matter.

$$_{24}C_5 = 42{,}504$$

Permutation and Combination

You probably learned to count before you even started school, but the method you learned (1, 2, 3, . . .) can be limiting. For example, how many seating arrangements are possible for your classroom? Don't answer that now. Just imagine how long it would take to list them all and count. Permutation and combination formulas help us to count things without actually having to see them.

PERMUTATION

Ms. Caba is giving a small prize to each of the top three students in her class. Each prize is different, and she is distributing them randomly among the winners. How many ways are there to distribute three prizes among three students? Each arrangement is a *permutation*.

1. Open **Permutation and Combination.gsp.** The first page, labeled Practice, has a set of icons grouped in a box. Each icon represents a different prize. The buttons allow you to adjust the number of prizes between zero and five.

2. Set the number of prizes to three. Select a prize and drag it out of the box. A copy of it will remain. Drag one of each prize and arrange them in a row.

3. Again, drag one of each prize out of the box and arrange them in another row, but this time in a different order. Continue forming rows of the three prizes until you have formed every possible arrangement.

Q1 How many arrangements were you able to form? In other words, what is the number of permutations for three distinct objects drawn from a set of three?

Q2 Without counting, how many arrangements do you think you could form with four prizes? (*Hint:* How many arrangements are there in which the first prize is a triangle?)

4. Open the Factorial page. Adjust the number of prizes to three, and press the *List Permutations* button. It will answer Q1 and display the possible arrangements, six in all.

Note that the permutations are listed in a logical order. For the first position there are three prizes to choose from. No matter what you choose for the first, there are two prizes available for the second position. That leaves only one remaining for the last position. Hence, the number of permutations is $3 \cdot 2 \cdot 1 = 6$. This number is written 3! (three factorial).

5. This page will list the number of permutations for any set of prizes numbering between zero and five. Try them all.

Q3 As you can see, the number of permutations gets out of hand in a hurry, which is why this demonstration has an upper limit. How many permutations are there for eight distinct objects drawn from a set of eight?

Q4 For any positive integer n you can calculate $n!$ by taking the product of all positive integers less than or equal to n. But what about zero? Why is $0! = 1$?

SUBSETS

In his class Mr. Brownlow has decided to give a prize to anyone who scores 100% on the final exam. He has wrapped four different prizes, but only two students have qualified. In this case you are still counting permutations, but you are not using up all of the available prizes.

Q5 Given a set of four prizes, you must choose two. How many prizes are available for the first student? How many are available for the second? What is the number of permutations for two objects chosen from a set of four?

The number of permutations for r objects chosen from a set of n is written $_nP_r$. You can compute it from the following formula:

You may also see this written as P_r^n or $P(n, r)$.

$$_nP_r = \frac{n!}{(n-r)!}, \quad \text{where } n \text{ and } r \text{ are integers and } n \geq r \geq 0$$

6. Go to the page labeled Permutation. This allows you to change the n and r parameters in a permutation calculation, again with an upper limit of five. List the permutations and check your answer to Q5. Experiment with other settings.

Q6 You may observe the fact that $_nP_0 = 1$ for any non-negative integer n. Explain why this is true.

Q7 It is also true that $_nP_n = {_nP_{n-1}}$. Explain why.

COMBINATION

Mr. Bozich promised that at the end of the semester he would give a prize to every student who scored 100% on any test. Stacey was the only student who accomplished this feat, and she did it three times. Mr. Bozich has five different prizes wrapped, and he tells Stacey to choose three.

In this case the order of the selections does not matter, because they are all going to the same student. Here you are not counting permutations; you are counting *combinations*.

7. Stay on the Permutation page for now. List the permutations for $_5P_3$. There should be 60.

Q8 The first permutation has a circle, a square, and a star. Looking carefully, you can see that six of the permutations have this same combination of prizes. Find them. Explain why there must be six.

You cannot count this same combination six times, and the same goes for all of the other combinations. Therefore, divide the number of permutations by six to get the number of combinations. The expression $_nC_r$ represents the number of possible combinations of r objects chosen from a set of n.

The combination may also be written C_r^n, $\binom{n}{r}$, or $C(n, r)$, and is often pronounced "n choose r."

$$_nC_r = \frac{_nP_r}{r!} = \frac{n!}{r!(n-r)!}, \quad \text{where } n \text{ and } r \text{ are integers and } n \geq r \geq 0$$

8. Go to the page labeled Combination. List $_5C_3$. Experiment with other combinations.

Q9 What are $_nC_0$, $_nC_1$, and $_nC_n$?

Q10 Explain why $_nC_r$ is always the same as $_nC_{n-r}$.

CALCULATIONS

9. Open the page labeled Calculations. This page has calculations for factorial, permutation, and combination. There are no graphical representations, but there are also no upper limits for the parameters.

Q11 A basketball team has ten players, and there are five different player positions on the floor. How many different starting lineups are possible?

Q12 Nine apartment tenants all drive, and their parking lot has only nine spaces. How many ways are there to arrange the cars in the spaces?

Q13 At the start of the game of euchre, each player gets five cards from a deck that includes only the cards from nine up to ace (9, 10, J, Q, K, A) in the usual four suits. How many hands are possible in euchre?

Fitting Functions to Data

MAKE THE SQUARES

Q1 The data appear to be periodic, so the $\sin x$ or $\cos x$ function may be a good choice for the parent function.

Q2 Slider b adjusts the horizontal stretch/shrink. Slider k adjusts the vertical translation.

5. If students are not familiar with the Custom Tools menu, show them how to press and hold the **Custom** tools icon to make it appear.

MAKE THE SQUARES SMALL (LEAST SQUARES)

Answers will vary for the questions asking students to actually fit functions. Some typical values for Q3, Q4, Q6, Q8, and Q10 are shown in the table below, but there is the possibility for considerable variation. For instance, the sine function used in Q3 will require very different values for the horizontal translation (h) if the vertical stretch (a) is negative instead of positive.

Q5 You need only two of the sliders for a linear function. The values of a and b combine to determine the slope. The values of h and k combine to determine the intercept.

Q7 You need only three of the sliders for a quadratic function. The values of a and b combine to determine the width of the parabola. (The values of h and k determine the vertex.) Similarly, square root and absolute value functions need only three sliders.

Q9 One reason for squaring the values is to make them all positive. If you just add the deviations, a large negative deviation and a large positive one might add up to zero, even though both points are far from the graph.

Q3 (page 1)	$y - 2.05\sin\left(\frac{x - 5.20}{1.45}\right) + 0.90$
Q4 (page 2)	$y = -0.60x + 2.40$
Q6 (page 3)	$y = -\left(\frac{x - 14.95}{3.20}\right)^2 + 25.25$
Q8 (page 4)	$y = 50.25 \cdot 2^{-(x - 0.25)/12.60} - 0.25$
Q10 (page 5)	$y = 1.25\sqrt{x - 7.90} + 0.15$
Q10 (page 6)	$y = \frac{7.5}{x - 5.00} + 1.05$

WHOLE-CLASS PRESENTATION

Use the Presenter Notes and **Fitting Functions Present.gsp** to present this activity to the whole class.

Fitting Functions to Data

Use this presentation to review function transformations and to introduce students to some of the principles of curve fitting.

MAKE THE SQUARES

1. Open **Fitting Functions Present.gsp.** The sketch has ten data points and four sliders. Tell students that their job is to transform a parent function using the sliders so that it fits the data points as closely as possible.

Q1 Looking at the pattern of these data points, what sort of parent function might you use? (The data look periodic, so $\sin x$ or $\cos x$ might be a good choice.)

2. Press *Show Parent Function* to reveal the parent function for this page.

3. Press *Show Transformed Function* to create and graph the transformed function.

$$g(x) = a \cdot f\!\left(\frac{x - h}{b}\right) + k$$

It's best to have a student operating the computer and taking direction from you and from other students.

Q2 Ask students how to drag the sliders to experiment with the transformation. You're not trying to fit the data yet; you are familiarizing students with the effects of the sliders. Students should drive this adjustment process. Don't rush them; give them time to discuss, to argue, and to think about what's happening.

Q3 Which slider adjusts the horizontal stretch/shrink? (b)

Q4 Which slider adjusts the vertical translation? (k)

4. Adjust the sliders to arrange the transformed function plot in a rough approximation of the data.

Now you're ready to measure the distances and find the sum of the squares.

5. Use the **Initialize Function and Sum, Next Square,** and **Show Result** custom tools to find the sum of the squares of the deviations. Consult the directions for the student activity for details on using these tools.

To make very small slider adjustments, select the point on the slider and press the right or left arrow key on the keyboard.

Q5 Adjust the sliders to minimize the sum of the squares. Switch back and forth among the various sliders to be sure you've found a minimum result. Record the slider values.

Q6 Why do you think we square the distances from the point to the function before adding them up? Why don't we just find the sum of the distances and use that to measure how good the fit is? (We use squared values to make all the values positive so that positive and negative deviations do not cancel each other out.)

The remaining pages contain a number of interesting data patterns to fit.

Fitting Functions to Data

The real world doesn't often behave as cleanly as mathematical functions do. The precision and consistency of pure mathematics is one of its attractions, but also one of its challenges: It's not easy to fit the messy data of real life into the more orderly world of mathematics. In this activity you'll use a general technique for fitting functions (which are smooth and well defined) to data (which may not be).

The technique involves starting with a parent function and using translations, stretches, and shrinks to generate a transformed function that more closely fits the data. To get the best fit, you need to be able to measure how close the fit is. You will use the method of *least squares* for this purpose: You will find the vertical distances from the data points to the function, square those distances, and add them up. The smaller the sum, the better the fit. (This is the reason for the method's name.)

MAKE THE SQUARES

1. Open **Fitting Functions.gsp.**

The sketch has ten data points and four sliders. Your job is to transform a parent function using the sliders so that it fits the data points as closely as possible.

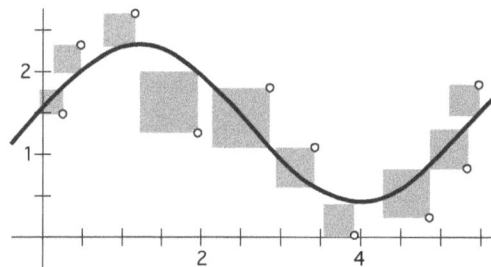

Q1 Looking at the pattern of these data points, what parent function might you use?

2. Press *Show Parent Function* to reveal the parent function to use on this page.

To graph g(x), choose **Graph | Plot New Function.** *To enter a slider value into the formula, click it in the sketch.*

3. Use the parent function and the sliders to graph the transformed function

$$g(x) = a \cdot f\left(\frac{x - h}{b}\right) + k$$

Q2 Drag the sliders to change the transformation. Which slider adjusts the horizontal stretch/shrink? Which one adjusts the vertical translation?

4. Arrange the transformed function plot in a rough approximation of the data.

Now you're ready to measure the distances and find the sum of the squares.

You cannot see the sum yet because it is hidden.

5. Press and hold the **Custom** tools icon and choose **Initialize Function and Sum** from the menu that appears. To start the summation process, click this tool on the transformed function. (Click the function itself, not the graph.)

Press and hold the **Custom** *tools icon to see the menu again.*

6. To construct the square for the first data point, choose the **Next Square** custom tool from the Custom Tools menu, and click it on the point.

7. Construct squares for each of the remaining data points.

8. To see the sum of the squares, choose the **Show Result** custom tool. You don't even have to click this tool; it shows the sum as soon as you choose it from the menu.

MAKE THE SQUARES SMALL (LEAST SQUARES)

As you adjust, you'll need to switch back and forth among the different sliders to get the best fit.

Q3 Drag the sliders while you watch the *Sum of Squares* calculation, and try to make this sum as small as possible. When you're satisfied, record the slider values and the *Sum of Squares* that you used.

Q4 Page 2 contains data to fit with a linear function. Transform the parent function, $f(x) = x$. Then record the slider values and the sum of squares that gives the best fit.

Q5 How many sliders do you really need in order to adjust a linear function? Explain why you don't need all four transformations in this case.

To make very small slider adjustments, select the point on the slider and press the right or left arrow key on the keyboard.

Q6 Page 3 contains data to fit with a quadratic function. The parent function is $f(x) = x^2$. Chris adjusted the sliders to make the sum of the squares 12.10. Can you do better? Record the slider values and the sum of squares that gives the best fit.

Q7 How many sliders do you really need in order to adjust a quadratic function? Explain why you don't need all four transformations in this case.

Q8 Page 4 contains yet more data and an exponential parent function. Fit the transformed function to the data and record your results.

Q9 Why do you think we square the distances from the point to the function before adding them up? Why don't we just find the sum of the distances and use that to measure how good the fit is?

EXPLORE MORE

Q10 Pages 5 and 6 contain additional data, but no functions. On these pages you must create your own parent function, transform it, and fit it to the data points. Choose your function carefully; for some data you may want to try two different parent functions to see which one fits best. (To do a second least-squares calculation in the same sketch, start over with the **Initialize Function and Sum** tool, and then use the **Next Square** tool on each data point.)

9. Collect some data of your own. The data can come from your own measurements, from a science lab, from the Internet, or from some other source. Plot the data in Sketchpad, and then choose an appropriate function and fit it to the data. Present the results to your group or to your entire class.

www.ingramcontent.com/pod-product-compliance
Lightning Source LLC
Chambersburg PA
CBHW080919220326
41598CB00034B/5616